Disaster Management

Concepts and Approaches

Disaster Management

Concepts and Approaches

Debabrata Mondal
MSc (Ag), PhD (Agril. Extension)

Mahesh High School
Hooghly, West Bengal

Debabrata Basu
MSc (Ag), PhD (Agril. Extension), Diploma in Distance Education

Professor and Head
Department of Agricultural Extension
Bidhan Chandra Krishi Viswavidyalaya
(Agricultural University)
Nadia, West Bengal

Oxford & IBH Publishing Co. Pvt. Ltd.
New Delhi
(*A Unit of* CBS Publishers & Distributors Pvt Ltd)

CBS Publishers & Distributors Pvt Ltd

New Delhi • Bengaluru • Chennai • Kochi • Kolkata • Mumbai
Bhopal • Bhubaneswar • Hyderabad • Jharkhand • Nagpur
• Patna • Pune • Uttarakhand • Dhaka (Bangladesh) • Kathmandu (Nepal)

Disaster Management
Concept and Approaches

ISBN: 978-93-89396-25-6

Copyright © Authors and Publisher

First Edition: 2020

OXFORD & IBH
New Delhi
(*A Unit of* CBS Publishers & Distributors Pvt Ltd)

Published by Satish Kumar Jain and produced by Varun Jain for
Oxford & IBH Publishing Co. Pvt. Ltd. under an imprint of
CBS Publishers & Distributors Pvt Ltd
4819/XI Prahlad Street, 24 Ansari Road, Daryaganj, New Delhi 110 002, India.
Ph: 23289259, 23266861, 23266867   Fax: 011-23243014   Website: www.cbspd.com
e-mail: delhi@cbspd.com; cbspubs@airtelmail.in.

Corporate Office: 204 FIE, Industrial Area, Patparganj, Delhi 110 092
Ph: 011-4934 4934   Fax: 011-4934 4935   e-mail: publishing@cbspd.com;   publicity@cbspd.com

Branches

- **Bengaluru:** Seema House 2975, 17th Cross, K.R. Road, Banasankari 2nd Stage, Bengaluru 560 070, Karnataka
  Ph: +91-80-26771678/79   Fax: +91-80-26771680   e-mail: bangalore@cbspd.com
- **Chennai:** 7, Subbaraya Street, Shenoy Nagar, Chennai 600 030, Tamil Nadu
  Ph: +91-44-26260666, 26208620   Fax: +91-44-42032115   e-mail: chennai@cbspd.com
- **Kochi:** 42/1325, 1326, Power House Road, Opp KSEB Power House, Ernakulam 682 018, Kochi, Kerala
  Ph: +91-484-4059061-65   Fax: +91-484-4059065   e-mail: kochi@cbspd.com
- **Kolkata:** No. 6/B, Ground Floor, Rameswar Shaw Road, Kolkata-700014 (West Bengal), India
  Ph: +91-33-2289-1126, 2289-1127, 2289-1128   e-mail: kolkata@cbspd.com
- **Mumbai:** 83-C, Dr E Moses Road, Worli, Mumbai-400018, Maharashtra
  Ph: +91-22-24902340/41   Fax: +91-22-24902342   e-mail: mumbai@cbspd.com

Representatives

Bhopal	0-8319310552	Bhubaneswar	0-9911037372	Hyderabad	0-9885175004
Jharkhand	0-9811541605	Nagpur	0-9421945513	Patna	0-9334159340
Pune	0-9623451994	Uttarakhand	0-9716462459		
Dhaka (Bangladesh)	01912-003485	Kathmandu (Nepal)	977-9818742655		

Printed at Glorious Printers, Daryaganj, Delhi, India

Preface

Over the last three decades natural and man-made disasters have been increasing day by day in terms of frequency, size, number of people affected, and material damage caused. Natural hazards are not avoidable, but they do not necessarily become natural disasters. The main thing put up to society is the increase of the vulnerability of human communities and infrastructures. If any previous major event happened today, it would cause much more consequences than it did at the time. This is a result of a rapid urbanisation process, increase population and infrastructure, economic developments, expansion of real-time communication, industrial interdependence that make the system more sensitive to disaster impact. The most affected communities and areas are those where lack proper planning and environmental practices were implemented in due time.

India has been actively pursuing a paradigm shift from relief centric approach in the past to the recent holistic way, encompassing all facets of disaster management. Ministry of Home Affairs, Government of India, is the nodal agency responsible for coordination of disaster management at national level. Earlier, government has been focusing on building capacity for preparedness and prevention. To achieve sustainable development, goals are not possible unless development is disaster-resilient. This requires appropriate institutional arrangements, building capacities at all the levels to carry out disaster risk reduction measures, and developing appropriate tools and guidelines.

Disaster management is a mandatory course for all the master's degree students of agricultural extension across all the agricultural universities and also it is an integral part of the mandatory environmental science course across all the general degree colleges and universities.

Nowadays, disaster management is being also included in the courses and instructions provided during the induction training towards civil service and alike. Not only that but also disaster management course is taught in multifarious dimensions of management studies, distance and e-learning and so on.

This book has been designed keeping in mind the new syllabus of the ICAR's agriculture course curriculum for postgraduate students. It covers different types of natural and man-made disasters, effects of various types of disasters, and their dimensions, characteristics and mitigation measures. It outlines the role of national, state, district and local governments, as well as the United Nations, in complex emergencies in prevention and control of disasters. Equally important is the knowledge about various international and national agencies involved in disaster relief and humanitarian assistance. Various approaches and theories have been described at a functional level. The book presents a set of principles and a model of the disaster management that are most suitable under the Indian conditions.

Debabrata Mondal
Debabrata Basu

Contents

Introduction

We cannot stop natural disasters but we can arm ourselves with knowledge: so many lives wouldn't have to be lost if there was enough disaster preparedness.

— Petra Nemcova

India, a large country with a geographical area of 3.28 million sq km, has a tropical and subtropical climate and is bounded in the north by the Himalayan mountain ranges. The wide Indo-Gangetic plain lies between the Himalayas in the north and the Deccan Plateau that occupies most parts of the Southern peninsular India. The Western and the Eastern Ghats constitute long mountain ranges, running along the west and the east coast of the peninsula. These Ghats leave narrow stretches of coastal plains along the Arabian Sea on the west and wider plains on the Bay of Bengal coast on the east. The country receives an annual precipitation of 400 million hectare metres, 73% of which is received between June and September. The Asia Pacific Region faces over 60% of the world's natural disasters. India, on account of its geographical position, climate and geological setting, has had from time immemorial, a fair share of these disasters. There is hardly a year when some part of the country or other does not face the spectra of drought, due to the failure of monsoons in vulnerable areas. One or two cyclones strike the peninsular region of the country every year. Similarly, floods are a regular feature of the Eastern India where Himalayan rivers inundate large part of its catchment area uprooting people, disrupting livelihood and damaging infrastructure (GOI-UNDP, 2009).

Natural disasters are global phenomena, which may occur anywhere, anytime with or without any indication. These adversely affect the lives of the people and cause considerable damage to the property, environment and the infrastructure worldwide. Natural disasters are impediments to attain sustainable development. India is a major disaster prone country in Asia Pacific Region. India has been traditionally vulnerable to natural disasters on account of its unique geo-climatic conditions. Floods, droughts, cyclones, earthquakes and landslides have been recurrent phenomenon. As per the natural hazards map of India, 60% of the landmass is prone to earthquakes of various intensities; an area of over 40 million hectares is prone to floods; about 8% of the total area is prone to cyclones and 68% of the area is susceptible to drought. In the decade of 1990–2000, an average of about 4344 people lost their lives and about a million people were affected by disasters each year. The loss in terms of private, community and public assets has been astronomical. Almost 85% of the country is vulnerable to single or multiple disasters. Of the 35 states and union territories in the country, 27 are disaster prone. The 229 districts of India are prone to multiple hazards, West Bengal for example is prone to four types of hazards. Floods, droughts, earthquakes, cyclones, landslides and avalanches have taken a heavy toll of lives and have caused enormous damage to property. Thirteen coastal states and Union Territories (UTs) in the country, encompassing 84 coastal districts, are affected by tropical cyclones. Four states

(Tamil Nadu, Andhra Pradesh, Orissa and West Bengal) and one UT (Puducherry) on the east coast and one state (Gujarat) on the west coast are more vulnerable to hazards associated with cyclones.

Sustainable approach in disaster mitigation gradually developed among the rural people, sustainability must aim at ensuring the protection, conservation and better management of natural resources. The main aim of sustainability is to mitigate the conflict between development and environment which indirectly increases disaster as people and environment interaction cannot be avoided. In fact, it is only the sustainable approach that ensures the negative effects to be minimum. To mitigate the effects of natural calamities, short-term strategies like relief and long-term programmes such as avenue plantations, construction of major and medium projects, soil and water conservation measures may serve to minimise flood and cyclone. Non-government organisations and the agencies associated with the activities like relief, employment, rehabilitation, technology generation and dissemination are to be included as an active partner to improve the situation in flood and cyclone prone areas.

Systematic integration of disaster risk reduction into development planning and programming at all scales is necessary. A paradigm shift in disaster management from conventional response and relied practice to a more comprehensive risk reduction culture is slowly taking place; however more needs to be done in the sector. Local people need to increase their awareness on both pre-disaster risk identification and reduction and post-disaster reconstruction process. They need to be involved in identifying the risk of embankment breach and of drainage congestion. Moreover, involvement of locals in post-disaster will give them a chance both to learn to handle such situations and earn a livelihood.

Technical capacity must be developed to manage risks and disasters within the government system at all levels including having courses at university level to strengthen the flow of information between scientific institution and practitioners. Therefore, the results of the study will be helpful in analyzing the many aspects for management of disaster.

Disaster Management in Agriculture

Disasters play havoc with loss lives, damage to property and infrastructure. There are many reasons of occurrence of disasters. These are increase in population; modernization, industrialization, depletion of resources, lack of knowledge and skills and rising economic disparities are considered to be mainly responsible for the vulnerability of communities to this disaster. The situation, though more or less similar globally, is quite distressing in countries of South Asia, such as India.

Agriculture depends on climate, as weather and climate are the primary factors in agricultural production. Increasing temperatures and varied rainfall due to emission of high level of carbon dioxide, both which will greatly impact the agricultural sector.

One of the major effects of disasters have, is on agriculture. A large portion of the population depends on agriculture for their livelihood. Agriculture is adversely affected by any abnormal weather changes or variations in physical conditions. Crop losses account about 35% of the world crop production due to pest, diseases, animals, insects and others. This invites cyclones, floods and droughts resulting in disruption of life and destruction to physical properties and loss of livelihood and stress of disasters.

Mondal et al (2014) showed that before the disaster the majority of people were self-sufficient and largely depended on agriculture for their livelihood. After the disaster major portion of workforce shifted their occupation agriculture to daily wages. The others continued their occupation remain more or less same. Following the disaster, many farmers and fishermen were left jobless as most of the agriculture land was inundated with saline water. Agriculture land was inundated and caused huge damage to crop. It is evident from the study that major portions of the workforce engaged to daily wages labourers. However, Mahatma Gandhi National Rural Employment Programme has created some scope for creating additional employment to the job—seekers at panchayat level. People were migrated to nearby town or cities for searching of jobs, as the only alternative for livelihood earning. Migration has highly increased after this disaster. These results most of the cultivators can cultivate only one crop a year. Many have become wage labours and have left for other places in India in search of jobs. The women folk, back at home, have to do all the work on field and home. Children hardly find time to go to school. Aged does not want to leave their ancestral home even if in dilapidated condition and susceptible to further hazards.

The importance of disaster management in agriculture was perceived and its additionally fortifying through an exhaustive disaster management policy by the Government of India giving anticipatory preparedness, mitigation, prevention and recovery was unequivocally suggested.

State government should be more proactive in dealing with disaster related issues prompting powerful administration in farming.

It was emphatically felt that notwithstanding the specialised and managerial skills in connection to disaster and related exercises significance ought to be given to disaster management education and training to the students on disaster preparedness, relief and recovery; make mindfulness about compelling calamity reaction in different crisis situation; acquainted with the students with devices for meeting medical requirement; incorporate gender sensitive, sympathy-based disaster management approach; and instill new aptitudes and hone the current abilities of government authorities, experts and delegates for effective management.

The main roles of education and training on disaster management have been emphasized in planning and implementation of disaster management strategies. The researchers recommended that the education should be designed to provide comprehensive knowledge on different types of hazards, disaster management techniques and impediments in the way of disaster reduction and should address to the community needs.

State agricultural universities should introduce, develop and launch a course on disaster management through multidisciplinary approach and National Academy of Agricultural Science to be the nodal office for disaster management in agriculture with special emphasis should be given on research, education and training (NAAS, 2004).

Disaster management in agriculture educational program must be developed and ensured to build up joint guidelines of training the country over. This is important to ensure consistency in norms of compassionate and mitigation measures in the area.

Academicians, scientists, researchers in the universities or research institutions, who are working in disaster management, ought to be associated with changing over a calamity into another open door with enhanced innovation sources of information and application, and they ought to be energized for their dynamic investment particularly in a disaster management in the field of agriculture.

There is need to strengthen research system to develop continuous analysis, activity/reflection which will translate the holes in a disaster management plans. Towards this end, the organisations/colleges might be urged to attempt the multidisciplinary research projects.

Advancement and protection of occupation in farming and development of alternative livelihood strategies, were accentuated to enhance the livelihood of rural people.

The effect of agrarian practices on condition clearing path for environmental calamity was related with respect to disaster management in agriculture.

Capabilities of PRI functionaries should be enhanced through education and training programmes in the field of agriculture and disaster management, which will help to develop well-trained self-help groups and community-based organisations.

Agriculture education on various tools and techniques to combat calamities will surely help the agricultural output. Further, contingency agricultural planning for the disaster prone can be of greater help to farmers living in those regions. Various organisations, institutions also provide education on the issue.

Natural Disaster

Disaster

The word 'Disaster' derived from Italian *disastro*, disaster; originally meaning "unfavourable to one's stars", from *dis-Italian, dis-*, bad (compare dys-), + *astro*, star, celestial body, from Latin *astrum*, from Greek *astron*.) is the impact of natural or man-made hazards that negatively affects society or environment.

"Disaster" means a catastrophe or mishap; a grave occurrence affecting any area; arising from nature or man-made cause, or by accident or negligence which results in substantial loss of life of human suffering, or damage to, and destruction of property, or damage to, or degradation of environment, and in of such a nature or magnitude as to be beyond the copying capacity of the community of the affected area (Govt. of India, 2005). Hence, a disaster is said to be take place when it includes two elements, namely hazard and vulnerability.

Hazards

Hazards are defined as the phenomena that pose a threat to people, structures, environmental resources and economic assets and which may cause a disaster. They could be either man-made or naturally occurring hazards in our environment.

Vulnerability

Vulnerability is defined as the extent to which a community, structure, service, or geographic area is likely to be damaged or disrupted by the impact of particular hazard, on account of their nature, construction and proximity to hazardous terrain or disaster prone area.

The relationship between hazard and vulnerability is shown in Fig. 1.1.

The relationship can be written as:

Disaster risk = Hazard + Vulnerability–Capacity.

Natural disasters are global phenomena, which may occur anywhere, anytime with or without any indication. These adversely affect the lives of the people and causes considerable damage to the property, environment and the infrastructure worldwide. Natural disasters are impediments to attain sustainable development. India is major disaster prone country in Asia-Pacific region.

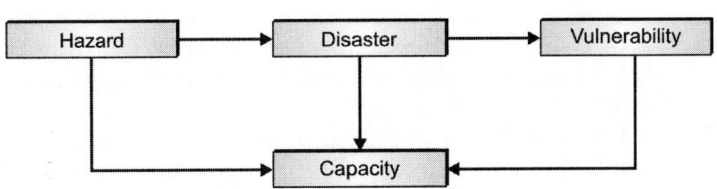

Fig: 1.1: Relationship between hazard and vulnerability

The Indian sub-continent is vulnerable to drought, floods, cyclones and earthquakes. Among the 32 States/Union domains in the nation 22 are multi-calamity inclined. As much as 40 million hectares of land in nation has been recognized as flood prone, and on a normal 18.6 million hectare of land is overwhelmed every year. Around 57% of zone of the nation is powerless against seismic action. 18% of nation's aggregate territory is drought prone, roughly 50 million individuals are yearly influenced by dry season and around 68% of the aggregate sown zone of the nation is dry spell inclined.

A **natural disaster** is an event resulting from natural processes of the earth such as floods, cyclone, drought, volcanic eruptions, earthquakes, etc. It can cause loss of life or destruction of property, economic, social and environmental losses.

The meaning of natural disaster is any calamitous situation that is caused by nature or the characteristic procedures of the earth. The seriousness of a disaster is estimated in lives lost, financial misfortune, and the capacity of the populace to remake. Occasions that happen in uninhabited territories are not thought about calamities. So, a surge on a uninhabited island would not consider a catastrophe, but rather a surge in a populated region is known as a cataclysmic event.

Researchers, geologists, and scientists endeavour to anticipate major disasters and deflect however much harm as could reasonably be expected. With all the innovation accessible, it is turned out to be simpler to foresee real storms, blizzards, cyclones, and other climate related catastrophic events. In any case, there are as yet cataclysmic events that surface rather out of the blue, for example, seismic tremors, out of control fires, avalanches, or even volcanic ejections. Some of the time, a period of caution is there, however, it is frequently short with cataclysmic outcomes. Territories that are not used to catastrophes influenced by streak surges or sudden hail tempests can be influenced in an outrageous way.

NATURAL DISASTERS AND THEIR MAIN CHARACTERISTICS

The disasters can also be categorised under two other methods:

i. On the basis of nature and causative factors such as natural and man-made disasters; and

ii. Occurrence time such as slow-on setting and quick on setting disasters.

The slow-onset disasters are those which take some time to create their impact and thus provide sufficient response time. The quick-onset disasters occur so quickly that there is very little time to respond and in some cases they just occur without providing any response time. Some of the quick-onset disasters can be predicted and thus forecasting and early warning could be passed on to the people in the threat zone, while in some cases early warning is not possible. Thus, the quick-onset disasters can be further divided into two subcategories:

i. Those which can be predicted with greater precision and provide some response time; and

ii. Those which cannot be predicted with precision and occur suddenly surprising every one and provide no response time.

Impacts of Disasters

Each types of disaster can have various problematic impacts. These thus cause for the most part unsurprising issues and needs of four sorts: (i) Ecological; (ii) well-being; (iii) social, financial, and political; and (iv) regulatory and administrative.

Ecological Effects

Debacles can have any number or mix of four impacts: (i) Obliteration and harm to homes and structures; (ii) diminished amount or nature of water supplies; (iii) devastation of yields and additionally sustenance stocks; and (iv) the nearness of unburied human bodies or creature bodies. These ecological impacts shift extensively from calamity to catastrophe. For instance, seismic tremors influence structures however as a rule not crops, while tropical typhoons may influence both. Firmly identified with the ecological impact is the effect that fiascos have ashore residency and qualities. These impacts likewise differ with the catastrophe compose; for instance, arrive values after seismic tremors will go up in zones that were not vigorously harmed, but rather arrive values go down in zones of dynamic volcanoes.

Consequences for Health

Sudden cataclysmic events are frequently accepted to cause far reaching passing as well as gigantic social interruption and episodes of plague illness and starvation, leaving survivors totally subject to outside alleviation. Orderly perception of the impacts of calamity on human well-being has prompted rather unique conclusions, both about the impacts of debacle on well-being and about the best methods for giving alleviation. In spite of the fact that all calamities are novel in that they influence territories with contrasting social, restorative, and monetary foundations, there are still similarities between debacles that, if perceived, can enhance the administration of well-being alleviation and utilization of assets. The accompanying focuses might be noted:

- There is a connection between the sort of calamity and its impact on well-being. This is especially valid for the quick effect in causing wounds: Quakes frequently cause numerous wounds requiring restorative care, while surges, storm surges and seismic ocean waves cause generally few.
- Some impacts are a potential instead of an inescapable danger to well-being. For instance, populace development and other natural changes may prompt expanded danger of illness transmission, in spite of the fact that plagues by and large do not come about because of calamities.
- The real and potential well-being dangers after fiasco do not all happen in the meantime. Rather, they have a tendency to emerge at various circumstances and to

shift in significance inside a fiasco influenced territory. Along these lines, setbacks happen basically at the time and place of effect and require prompt restorative care. The dangers of expanded ailment transmission take more time to create and are most prominent where there is swarming and decreased benchmarks of sanitation.

- Disaster-made requirements for sustenance, safe house, and essential medicinal services are normally not add up to. Indeed, even uprooted people frequently rescue a portion of the essential necessities of life. Further, individuals by and large recuperate rapidly from their prompt stun and suddenly take part in hunt and protect, transport of the harmed, and other private help exercises.

Financial, Social, and Political Effects

Calamities disturb as opposed to crush economies. Amid a crisis, individuals must leave their occupations and dedicate their opportunity to fiasco related exercises, for example, inquiry and protect, or to care of survivors. Amid this period typical financial exercises are extremely diminished, regardless of whether the wellsprings of business are unaffected by the catastrophe. This period is fleeting, be that as it may, and in the later periods of a catastrophe financial exercises rapidly expect a high need for the two organisations and casualties alike. Regardless of whether an economy can recoup rapidly relies upon the misfortunes managed. Physical harm to organisations and industry may incidentally end a few exercises, yet most endeavours can work at lessened levels even with the loss of hardware. Frequently the specialists in a harmed manufacturing plant can be given something to do repairing or remake the office. Regardless, the loss of occupations is typically just impermanent.

A few spectators have noticed that blast economies regularly create after an across the board calamity, for example, a seismic tremor or sea tempest requiring major physical reproduction. Long haul impacts are not yet known, but rather no less than one exa-mination demonstrates that if low-wage casualties are given need in work contracting, blast economies can be a methods for changing a portion of the misfortunes.

Victims and Survivors

Almost everyone in the population is affected by a disaster. No one is untouched by it. Those who suffer damage are called victims. The victims may die or live. Those who manage to live are called survivors.

Psychosocial Aspects of Disaster

Often, minimal importance is given to the mental trauma suffered by the victims of a disaster. They are overshed owed by the excessive importance to physical and financial needs of the victims which are considered by the relief personnel to be more than sufficient to alleviate the suffering of the victims. Unlike physical and material damage, the damage to the psyche (mind) cannot be obviously seen, until and unless, it is looked for. And, to look for, the relief personnel need to be aware of the possible effects on the mind, which can be permanent and disabling. The psychosocial needs are generally seen as something too secondary to attract the attentions of relief agencies, relief workers and governmental organisations.

Concept of Loss due to Disaster

It has been of repeated mistake to assess severity of a disaster by means of calculating the loss in terms of numbers, quantity figures or units such as number of deaths, number of wounded person, number of houses damaged, surface area of affected land, etc. But, this is not the actual measure. The meaning of the loss rather than loss itself is a much more significant measure. In other words, the impact of the disaster itself is more important. For example, the loss of a neighbour may mean a great loss to one person but a minimal loss to another. Similarly, loss of animal life may mean nothing for one victim but may mean a lot for

an animal lover. The loss of a house may mean less for someone who is thankful for having survived, but more for someone who has a sentimental attachment to his house. Thus, the actual damage being less, the impact may be disproportionately severe. Though the entire population experiences the same disaster, each one perceives it in a different and unique way.

Phases of Traumatic Stress Reactions in a Disaster

Disasters and terrorist attacks are often widespread with many people who directly experience the event and many more who may witness or be indirectly impacted. Many people may encounter behavioural and emotional readjustment problems. Many post-traumatic stress symptoms are normal responses to overwhelming stressors. Expo-sure to these overwhelming stressors may change our assumptions about life and create distress, but the intensity of this distress will subside with time. Experts agree that the amount of time it takes people to recover depends both on what happened to them and on what meaning they give to those events.

Impact Phase

Most people respond appropriately during the impact of a disaster and react to protect their own lives and the lives of others. This is a natural and basic reaction. A range of such behaviours can occur, and these may also need to be dealt with and understood in the post-disaster period. After the fact, people may judge their actions during the disaster as not having fulfilled their own or others' expectations of themselves.

During the impact phase, some people respond in a way that is disorganised and stunned, and they may not be able to respond appropriately to protect themselves. Such disorganised or apathetic behaviour may be transient or may extend into the post-disaster period, so that people may be found wandering helpless in the devastation afterwards. These reactions may reflect cognitive distortions in response to the severe disaster stressors and may for some indicate a level of dissociation.

Several stressors may occur during impact, which may subsequently have consequences for the person:

- Threat to life and encounter with death
- Feelings of helplessness and powerlessness
- Loss (e.g. loved ones, home, possessions)
- Dislocation (i.e. separation from loved ones, home, familiar settings, neighbourhood, community)
- Feeling responsible (e.g. feeling as though could have done more)
- Inescapable horror (e.g. being trapped or tortured)
- Human malevolence (it is particularly difficult to cope with a disaster if it is seen as the result of deliberate human actions.)

Immediate Post-disaster Phase: Recoil and Rescue

This is the phase where there is recoil from the impact and the initial rescue activities commence. Initial mental health effects may appear (e.g. people show confusion, are stunned, or demonstrate high anxiety levels). Emotional reactions will be variable and depend on the individual's perceptions and experience of the different stressor elements noted earlier. Necessary activities of the rescue phase may delay these reactions, and they may appear more as the recovery processes get under way.

Conversely, relief and survival may lead to feelings of elation, which may be difficult to accept in the face of the destruction the disaster has wrought.

Recovery Phase

The recovery phase is the prolonged period of adjustment or return to equilibrium that the community and individuals must go through. It commences as rescue is completed and individuals and communities face the task of bringing their lives and activities back to normal. Much will depend on the extent of devastation and destruction that has occurred

as well as injuries and lives lost (Raphael, 1993).

Regulatory and Managerial Effects

Regulatory issues in calamity are made more troublesome by four components, which increment in significance with the degree of the catastrophe.

1. Effects on group initiative. The loss of pioneers because of death or damage can hinder calamity reaction.
2. Disruption of formal associations. At the point when a calamity strikes, extensive formal associations are generally disturbed. Little, people group based associations are by and large better ready to work, even with loss of pioneers.
3. Damage to basic offices and helps. Far reaching debacles can demolish or harm offices that might be basic for reacting to the catastrophe as well as for keeping up a sheltered domain and open request. Among these are correspondences establishments; electrical producing and transmission offices; water stockpiling, filtration, and pumping offices; sewage treatment offices; doctor's facilities; police headquarters; and other private structures.
4. Disruption of transportation (and detachment of assets). Amid the underlying phases of most sorts of calamities, every surface mean of transportation inside a group are disturbed. Scaffolds can be thumped out; streets can be cut via avalanches; rubble can square avenues and thru-ways.

Impacts on Agriculture

- The farming segment—including crops, domesticated animals, fisheries and ranger service—retains roughly 22% of the monetary effect caused by medium- and expansive scale common perils and calamities in creating nations;
- The high effect of normal dangers and calamities on agribusiness calls for improved mainstreaming of fiasco chance decrease and versatility working inside the farming parts;

- There are real information holes on the effect of common risks and calamities on the agribusiness areas in creating nations. This area particular information must be efficiently gathered and incorporated into national and global catastrophe misfortune databases to better advise proper hazard decrease arrangements and ventures for and inside the division;
- The farming areas should be activated as proactive usage accomplices for the conveyance of the post-2015 system on catastrophe hazard decrease in order to upgrade nearby activity and manufacture strength of the most defenseless, which are frequently additionally the most nourishment shaky.

Types of Disaster:

Disaster can be classified as follows:

1. According to nature/magnitude (Table 1.1)
2. According to timing and predictability (Table 1.2)
3. According to response time (Table 1.3)
4. According to impact (Table 1.4)

FLOODS

The inundation of an area by water is called a flood. In other words, when a river over flows its banks and water spreads in the surrounding areas is a flood. Of the annual rainfall, 75% is concentrated over four months of monsoon (June–September) and as a result almost all the rivers carry heavy discharge during this period. The flood hazards compounded by the problems of sediment deposition, drainage congestion and synchronization of river floods with sea tides in the coastal plains. The rivers originating in the Himalayas also carry a lot of sediment and cause erosion of the banks in the upper reaches and over-topping in the lower

segments. The most flood prone areas are the Brahmaputra and the basins in the Indo-Gangetic plains. The other flood prone areas are the north-west region of the west flowing rivers like Narmada and Tapi, the Central India and the Deccan region with major east

Table 1.1: Disaster according to nature/magnitude

Major natural disasters	Minor natural disasters
Flood	Cold wave
Cyclone	Snowfall
Drought	Thunderstorm
Earthquake	
Sunstroke	

Major man-made disasters	Minor man-made disasters
Setting of fire	Road and train accidents
Epidemic	Accidents during festivals
Deforestation	Food poisoning
Pollution due to excessive drink	Death due to spurious/ prawn culture
Chemical pollution	Industrial disaster/crisis
	Rehabilitation
	Acid rain
	Riots and war
	Environmental pollution

Table 1.2: Disaster according to timing and predictability

	Quick-onset	
Slow onset	Predictable	Unpredictable or sudden
Drought	Cyclone	Earthquake
Famine	Flood	Landslides
	Typhoon	Avalanches
	Heat or cold wave	Tsunami

Table 1.3: Disaster according to response time

Long response time	Short response time	No response time
Drought/Famine	Cyclone/Floods	Earthquake/Landslide

Table 1.4: Disaster according to impact

Affect all aspect of life	Loss of life and damage to physical infrastructure	Affect livelihood and cause environmental degradation	Threaten only lives
Cyclone/Tornado Flood Earthquake	Earthquake	Drought/Forest fire	Famine/epidemic

flowing rivers like Mahanadi, Krishna and Cavery. While the area liable to floods is 40 million hectares, the average area affected by floods annually is about 8 million hectares. The annual average cropped area affected is approximately 3.7 million hectares.

Various causes of flood are described below:

1. **Heavy rainfall:** Heavy rain in the catchment area of a river causes water to over flow its banks, which results in the flooding of nearby areas.

2. **Sediment deposition:** River beds become shallow due to sedimentation. The water carrying capacity of such river is reduced. As a result the heavy rains water over flows the river banks.

3. **Deforestation:** Vegetation hampers the flow of water and forces it to percolate in the ground. As a result of deforestation, the land becomes obstruction-free and water flows with greater speed into the rivers and causes flood.

4. **Cyclone:** Cyclone generated sea waves of abnormal height spreads the water in the adjoining coastal areas. In October 1994 Orissa cyclone and May, 2009 cyclone Aila in West Bengal generated severe floods and caused unprecedented loss of life and property.

5. **Interference in drainage system:** Drainage congestion caused by badly planned construction of bridges, roads, railway tracks, canals, etc. hampers the flow of water and the result is flood.

6. **Change in the course of the river:** Meanders and change in the course of the river cause floods.

7. **Tsunami:** Large coastal areas are flooded by rising sea water, when a tsunami strikes the coast.

Effects of Flood

Human lives: Human and domestic animals diminish because of suffocating, genuine wounds and episode of scourges like loose bowels, cholera, jaundice or viral contaminations are basic issues looked in surge influenced zones. Indeed, even wells, other wellspring of drinking water get submerged bringing about intense lack of safe drinking water amid surges. Therefore, regularly individuals are compelled to drink the tainted floodwater, which may cause genuine infections. For instance, in the recent floods in Kerala as many as 1.76 crore poultry, 46,000 cows and 20,000 goats were washed away (Fig. 1.2).

- **Structural loss:** Amid surges mud cabins and structures based on frail establishments fall imperiling human lives and property. Harm may likewise be cause to streets, rail, dams, landmarks, harvests and dairy cattle. Surges may remove trees and may cause avalanches and soil disintegration.

- **Material loss:** Family unit articles including eatables, electronic merchandise, beds, garments, furniture get submerged in water and get spoilt all materials mounted on ground, for example, sustenance stock, gear, vehicles, animals, apparatus, salt container and angling water crafts can be submerged and spoilt.

- **Utilities loss:** Utilities, for example, water supply, sewerage, correspondence lines, electrical cables, transportation system and railroads are put in danger.

- **Crop loss:** Aside from the loss of human and cows life, surges cause serious obliteration of standing rural harvests. Surges water ruins the put away nourishment grains or reaped trim. Surges may influence soil attributes and may turn them fruitless because of the disintegration of the best soil or in beach front regions farming grounds may turn saline because of flooding via ocean water.

Control Measures

Flood control can be accomplished through different means. The floodwater can be decreased by diminishing the run-off water through afforestation. Woodlands advance water permeation in the ground, in this manner reviving the groundwater and decreasing the run-off water. Development of dams likewise lessens surge water through capacity. Dams can store water, which cannot be suited in the stream, downstream may cause surges. Water can be discharged in a controlled way from the dam. Desalting, extending and expanding bank increment the limit of a stream/channel/deplete.

  i. Reservoirs: By building supplies in the courses of waterways could stores additional water at the season of surge. Such measures received till now be that as it may, have not been effective. Dams worked to control surges of Damodar could not control the surge.

 ii. Embankments: By building surge security dikes, surges water can be controlled from flooding the banks and spreading in close by zones. Working of dikes on Yamuna, close Delhi, has been fruitful in controlling the surge.

iii. Afforestation: The fuzzy of surge could be limited by planting trees in catchment regions of waterways.

 iv. Restoration of unique seepage framework: Drainage framework is for the most part stifled by the development of streets, channels railroad tracks and so on. Surges could be checked if the first type of waste framework is reestablished.

DROUGHT

The main cause of drought is lack of precipitation and specifically, the planning, circulation furthermore, power of this insufficiency in connection to existing stores. A drawn out time of generally dry climate

Fig. 1.2: Impact of Kerala flood, 2018

prompting dry season is a broadly perceived atmosphere irregularity. Dry season can be obliterating as water supplies go away, crops neglect to develop, creatures kick the bucket, and lack of healthy sustenance and sick well-being wind up across the board. The ecological impacts of dry season, including stalinization of soil also, groundwater decay, expanded contamination of freshwater biological communities and territorial eradication of creature species.

Main causes of drought are

1. Precipitation deficiency
2. Dry season
3. Erosion and human activities
4. Climatic changes

Mitigation Strategies for Drought

Dam: Construction of dams to supply water in the times of drought.

- *Cloud seeding:* Artificially created weather condition to induce rainfall.
- *Desalinization:* Use of seawater for consumption of crops.
- *Reduce erosion:* Crop rotation should be maintained properly and planted drought resistance crops.
- *Rainwater harvesting:* Storage of rain water from catchment.
- Building canals for irrigation in drought prone areas.

CYCLONES

Tropical cyclones are known around the world by various names: Hurricanes in the Atlantic and Caribbean, typhoons in the West Pacific, baguios in the Philippines, cordonazos in Mexico, tainos in Haiti. A tropical cyclone is essentially a rotating storm in the tropical oceans. It is conventionally defined as a circular storm with rotating wind speeds in excess of 64 knots (32 meters per second). The lifespan of a tropical cyclone is, on average, about 6–9 days until it enters land or recurves into temperate latitudes, but this may vary

from a few hours to as much as 3–4 weeks. Tropical cyclones form in the oceans between 5° and 30° north and south of the equator. They are found in all oceans of the world, with the probable exception of the South Atlantic and the South Pacific east of 140° west longitude.

Driving forces: Cyclones are created by intense low pressure areas in the atmosphere, due to which wind blows with a very high speed in circular motion. Horizontally, this area extends from 500 to 1000 km and vertically from surface to 12–14 km. There are two distinct cyclone seasons in Indian Subcontinent: Pre-monsoon (May–June) and post-monsoon (October–November) and are known for severe storms. The impact of these cyclones is confined to the coastal districts, the maximum destruction being within 100 km from the centre of cyclone and on either side of the storm track. Most casualties are caused due to coastal inundation by tidal waves, storm surges and torrential rains (Table 1.5).

Cyclones are born in the hot, humid late-summer environment of the tropics (June to August in the Caribbean, November to April in the South Pacific). As the sun warms the oceans, evaporation and conduction transfer heat to the atmosphere so rapidly that air and water temperatures seldom differ by more than 1°F. The water vapour generated by such evaporation is the fuel that drives a tropical storm, because as the vapour condenses into clouds and precipitation it pumps enormous amounts of heat into the cyclone. The fuel supply is controlled by the evaporation rate, which explains why cyclones cannot develop when the ocean temperature is below about 24°C (76°F).

The naming convention for cyclones in the Northern Indian Ocean started in 2004 when the India Meteorological Department (IMD) released a list of names that would serve to identify cyclonic storms. Eight members of ESCAP (Economic and Social Commission for Asia and Pacific)—Bangladesh, India, Maldives, Myanmar, Oman, Pakistan, Thailand and Sri Lanka, with coasts along the Arabian Sea or the Bay of Bengal—had

Table 1.5: Indian Classification of Cyclonic Disturbances in the Northern Indian Ocean (Bay of Bengal and Arabian Sea)

Type	Wind speed in km/hr	Wind speed in knots
Low pressure area	<31	<17
Depression	31–49	17–27
Deep depression	50–61	28–33
Cyclonic storms	62–88	34– 47
Severe cyclonic storm	89–118	48– 63
Very severe cyclonic storm	119– 221	64–119
Super cyclone	222 or more	120 or more

Source: Indian Meteorological Department, 2008

suggested four names each at the start of the programme.

The names are picked in sequence, one name after the other from each country in alphabetical order. The previous cyclone was named Bijli, Dr S.R. Ramanan, Director, Area Cyclone Warning Centre, Chennai, says: "We had submitted the first set of names to WMO (World Meteorological Organisation) in 2004. Aila was the 20th on the list and we have 12 left. The next one is going to be Phyan, coined by Myanmar. The next 32 have also been decided." Phyan is to be followed by Ward (Oman), Laila (Pakistan), Bandu (Sri Lanka) and Phet (Thailand). Even common people can suggest names of cyclones, says Dr. Ramanan. They can send their suggestions to the Director General of the Indian Meteorological Department.

There are certain criteria to coin a name for a cyclone. These are:

1. The name should be short and readily understand when broadcast.
2. The name must not be culturally sensitive.
3. As a storm brings death and destruction its name should not be used repeatedly.

The suggested name pertaining to India should be communicated to the Director General Meteorology, Indian Meteorological Department, Mausam Bhawan, New Delhi, for consideration.

Effects of Cyclones

Strong wind: The most common and may be best comprehended impact of cyclones is strong wind. Truth be told, these strong winds tend to influence the other dangerous operators of cyclones. Low-level winds will regularly be more grounded on the correct side of a tornado in the Northern Hemisphere, however the wind quality has a tendency to be exceedingly factor regardless of where a typhoon hits. The strong wind of tornadoes can cause harm over a zone of 25 km in littler frameworks and up to 500 km in bigger frameworks. Winds have been known to demolish littler structures and thump out power for a huge number of individuals.

The damage to the houses and other intra-structure due to cyclones seems to be very high, as large amount of water that inundated the village soaked the region and the mud and thatched houses sunk into the mud. People have also lost their land (Fig. 1.3).

Tornadoes

Tornadoes do not typically happen in the same tropical districts that violent winds normally influence; rather tornadoes for the most part originate from the tempests in waterfront areas and on islands. They might be much more typical than individuals once accepted. Violent wind generated tornadoes are regularly not revealed in districts, for example, the Caribbean, but rather some harm designs recommend that they happen every now and again. Tornadoes can achieve twist velocities of up to 480 kph and can extend in excess of 3 km. Typhoon tornadoes have a tendency to happen in the external edge of the eyewall

Fig. 1.3: Cyclone Aila, 2009 struck Bay of Bengal causing widespread damage, loss of life and rendering agricultural land barren from excess saline content

cloud, in the right-front quadrant of the tempest framework.

Tempest Surges

A tempest surge is an unusual ascent in water that happens amid a twister. Conceivably heartbreaking surges happen in seaside zones with low-lying territory that empowers immersion. The tempest surge is commonly the most harming impact of violent winds, verifiably bringing about 90% of tropical tornado passings. At the point when joined with solid breezes, storm surges can create gigantic waves that can cause inland flooding and demolition.

Precipitation and Flooding

The rainstorms delivered in a tornado framework create exceptional precipitation causing huge flooding, mudslides and avalanches. This flooding has a tendency to be more serious and damaging inland because of poor readiness. Despite the fact that this precipitation can be exceptionally dangerous and taken a toll a great many dollars in harm, rain in littler typhoon frameworks can really be valuable when it gives genuinely necessary precipitation to drier territories.

Erosion

A cyclone's high wind can disintegrate the soil, consequently harming existing vegetation and environments. This erosion leaves the zone presented and inclined to much more wind erosion. Soil and sand that is blown into different territories can harm the vegetation there.

Erosion likewise can be caused by storm surges from tropical violent winds. Waves that compass far onto a shoreline drag the sand over into the sea, leaving the influenced zone exceedingly dissolved. This can harm shoreline and ridge biological communities and also structures. The ocean will in the long run take the sand back to the shoreline, yet this can take years.

EARTHQUAKE

Earthquakes are one of the most dangerous and destructive forms of natural hazards. They strike with sudden impact and little warning. They may occur at any time of day or on any day of the year. An earthquake can devastate an entire city or a region of hundreds of square kilometers. They can reduce buildings to a pile of rubble in seconds, killing and injuring their inhabitants. Earthquakes are caused by the movement of massive land areas, called plates, on the earth's crust. Often covering areas larger than continents, these plates are in a constant state of motion. As the plates move relative to one another, stresses form and accumulate until a fracture or abrupt slippage occurs. This sudden release of stress is called an earthquake.

Earthquake is a series of underground shock waves and movements on the earth's surface. It is caused by natural processes wresting the earth's crust. It is found where one of the earth's plates is moving against another and building up so much tension that the rock cracks. The sudden cracks and the movement of the rocks send out shock waves (P-waves and S-waves) making the ground shake violently. The severity of an earthquake is measured on the Richter scale.

The Himalayan mountain ranges are considered to be the world's youngest fold mountain ranges. The subterranean Himalayan are, therefore, geologically very active. The Himalayan frontal arc, flanked by the Arakan Yoma belt in the east and the Chaman fault in the west constitute one of the most seismically active regions in the world. Four earthquakes exceeded the magnitude of 8 in the Richter scale in the history of last 60 years; these are the Assam earthquakes of 1950 and 1987, the Kangra earthquake of 1905 and the Bihar–Nepal earthquake of 1935. The peninsular part of India comprises stable continental crust regions which are considered stable as they are far away from the tectonic activity of the boundaries. Although these regions were considered seismically least active, an earthquake that occurred in Latur in Maharashtra on September 30, 1993 of magnitude 6.4 in the Richter scale caused substantial loss of lives and damage to infrastructure.

Effects of Earthquake

Tsunami: Tsunamis, which are famously and mistakenly, known as "tidal waves," are a grave risk to many parts of the world, especially around the Pacific Ocean basin. Tsunamis are a progression of water waves caused when the ocean bottom moves vertically in a tremor (which is the reason they are remarkable in California quakes—most CA seismic tremors are strike-slip, with next to zero vertical movement) and which can travel huge separations in a brief timeframe. Tidal wave speeds in the profound sea have been estimated at in excess of 700 km/hr, equivalent to some fly planes, and when waves achieve shallow water close to the drift, they can achieve statures of in excess of 27 meters (90 feet)! Keep in mind that torrents are a progression of waves, and may begin with a delicate withdrawal of water, trailed by an exceptionally unexpected arriving wave, trailed by another withdrawal, and so on. The most secure activity in the event that you hear a wave is coming is to move to higher ground far from the shoreline as fast as could be allowed.

Ground Shaking

Ground shaking is the most commonplace impact of earthquakes. It is an effect of the entry of seismic waves through the ground, and ranges from very delicate in little quakes to unimaginably savage in substantial tremors. In the 27 March 1964, Alaskan seismic tremor, for instance, solid ground shaking went on for as much as 7 minutes. Structures can be harmed or obliterated, individuals and creatures experience difficulty standing up or moving around, and articles can be hurled around because of solid ground shaking in quakes. In any case, you should take note of that, while numerous individuals are slaughtered in tremors, none are really executed straightforwardly by the shaking—on the off chance that you were out in an open field amid a greatness 9 quake, you would be to a great degree frightened (I know I would), however your shot of biting the dust would be zero or damn close it. It is simply because we endure in building structures, interstates, and so forth that individuals are murdered; it is our duty, not the earthquakes.

Ground Burst

Ground burst is another vital impact of quakes which happens when the seismic tremor development along blame really breaks the earth's surface. While dynamic ground break is similarly uncommon, there have been instances of it in California—for instance, amid the 1906 tremor, fences close Pt. Reyes were balanced by as much as 7 meters. What is more, in the Owens Valley seismic tremor in 1872, a blame scarp as much as 8 meters high broke the ground close Lone Pine. Break causes issues for people by, well, bursting things; pipelines, burrows, water channels, railroad lines, streets, and air terminal runways which cross a zone of dynamic crack can without much of a stretch be obliterated or seriously harmed.

Landslides

The stuns created by tremors especially in bumpy territories and mountains which are structurally touchy causes avalanches and flotsam and jetsam fall on human settlements and transport framework on the lower slant fragments, exacting harm to them.

Flash Floods

Strong seismic waves make harm dams accordingly causing serious blaze surges. Serious surges are additionally caused in view of obstructing of water stream of waterways because of shake squares and flotsam and jetsam created by extreme tremors in the slope inclines confronting the waterway valleys. In some cases, the blockage is severe to the point that streams change their principle course.

Fires

The solid vibrations caused by extreme quakes unequivocally shake the structures and along these lines causing serious flames in houses, mines and production lines as a result of upsetting of cooking gas, contact of live electric wires, stirring of impact heaters, dislodging of other fire related and electric apparatuses.

Deformation of Ground Surface

Extreme tremors and resultant vibrations caused by seismic tremors result in the twisting of ground surface due to rise and subsidence of ground surface and blaming movement (development of deficiencies).

Others

Liquefaction and subsidence of the ground are critical impacts which frequently are the reason for much annihilation in seismic tremors, especially in unconsolidated ground. Liquefaction is when residue grains are actually made to coast in groundwater, which makes the dirt lose all its strength. Subsidence would then be able to take after as the dirt recompacts. Sand blows, or sand volcanoes, frame when pressurised planes of groundwater get through the surface. They can shower mud and sand over a territory a couple of meters over. These impacts represent a grave peril to structures, streets, prepare lines, air terminal runways, gas lines, and so on. Structures have really tipped over and sunk mostly into condensed soils, as in the 1964 Niigata tremor in Japan. Underground gas tanks and septic tanks have been known to buoy to the surface through condensed soils. Everything considered, liquefaction and related impacts brought about more than $20 billion harm in the 1995 Kobe quake, and comparable levels of harm are conceivable in US port offices amid an extensive tremor.

Management and Mitigation Measures of Earthquakes

Earthquakes make colossal damage human life, condition, and environment. Human life requires prompt medicinal guide and recreation administrations. Be that as it may, we can be better arranged to meet tremors by:

i. Developing earthquake resistant building and structures.
ii. Seek shelter under stable tables or under door frames.
iii. After an earthquake, check gas, water and electricity pipes and lines for damage.
iv. Architects ought to take after the construction regulations that have been set around the Bureau of Indian Standards. The experts ought to guarantee that exclusive fitting designs are passed for another development.
v. Safeguard and alleviation tasks ought to be kept set up to meet all inevitabilities.
vi. Satisfactory restorative offices ought to be given to make minimum damage human life.
vii. Individuals ought to be made mindful of the significance of developing quake safe structures.
viii. Adequate preparedness and assistance in catastrophes is extremely important in areas affected by earthquakes.

LANDSLIDES

The Himalayan, the north-east hill ranges and the Western Ghats experience considerable landslide activities of varying intensities. The rocks and the debris carried by the rivers like Kosi originating in the Himalayas cause enormous landslides in the valleys. The seismic activity in the Himalayan region also results in considerable landslide movement. The heavy monsoon rainfall, often in association with cyclonic disturbances, results in considerable landslide activity on the slopes of the Western Ghats.

Main Causes of Landslides

Landslides are prompted by human exercises, for example, deforestation, explosive impacting of rocks, earth work, developments, vibrations from huge machines, and so forth.

The exercises that require chopping down of trees are chiefly considered to prompt avalanche in inclined territories. The trees work through their underlying foundations that hold the dirt set up.

- Slope instability because of expulsion of parallel and hidden support.
- Aimless slashing down of trees.
- Cut and consume development rehearses in slopes.
- Street development and mining exercises.
- With expanding populace weight, there is an expansion in touching exercises, urbanization which diminishes thick normal evergreen backwoods cover.
- Because of these exercises the biological adjust is disturbed, in this way bringing about releasing of the dirt.

Under states of overwhelming precipitation, there is expanded and significant soil disintegration and successive avalanches.

Impacts of Landslides in India

By and large landslides are activated by substantial or drawn out precipitation. Avalanches make extreme harm lives and property while additionally causing disturbance in correspondence systems and development of activity.

Consistently, landslides in the Himalayan locale execute individuals and make harm a few towns abandoning them unfit for home.

Landslides make barricades in the street organize and furthermore in stream framework, which causes surge. The terraced cultivate fields that are annihilated via landslides, cannot be effectively recuperated or made beneficial once more.

Influenced via landslides, the street organise stays shut for long stretches, thus, making tremendous hardships individuals possessing and reliant on the territory for their fundamental supplies and arrangements. Landslides disturb water sources and chocked them by garbage fall.

Because of landslides, the stream dregs stack is expanded significantly, which brings about sporadic courses of waterway and continuous breaking of banks likewise bringing about sudden surges. The water channels are likewise influenced because of interruption in past channels, this prompts unsettling influence in water supply to subordinate villagers for water system purposes. This at that point antagonistically influences agribusiness creation in the influenced area.

Mitigation Strategies of Landslides

Abundance water in catchment zones ought to be put away to diminish the impact of glimmer surges, this will likewise energize the groundwater level in regions inclined to avalanche in India.

The spillover gathering lakes in the catchment zones must be burrowed to store water.

On people group grounds, fuel or grain trees ought to be developed to build woodland cover to diminish avalanche danger in India.

- Frequency of landslides perils and sort of human action and also area decide affect.
- Total evasion of landslides peril territories or confinement on danger zone action is a compelling technique for administration.
- Land utilise arrangements and directions ought to likewise be set up in territories inclined to landslides.
- Hazard possibilities of destinations ought to be assessed.
- Landslides can be moderated in following ways:
  - The landslides can be secured with an impermeable layer
  - Surface water is coordinated far from the landslides
  - Groundwater is depleted from the landslides
  - Education and mindfulness about the effect of landslides is additionally an absolute necessity.

Eating ought to be limited and better grass must be developed at first glance already munched to expand the hang on soil by plant roots. These grasses can be of some business significance with the goal that financial returns energize ranchers in zones inclined to avalanche in India.

AVALANCHES

Avalanches constitute a major hazard in the higher reaches of the Himalayas. Heavy loss of life and property has been reported due to avalanches. Parts of the Himalayas receive snowfall round the year and adventure sports are in abundance in such locations. Severe snow avalanches are observed during and after snowfalls in Jammu & Kashmir, Himachal Pradesh and the hills of Western Uttar Pradesh. The population of about 20,000 in Nubra and Shyok valley and mountaineers and trekkers faces avalanche hazard on account of steep fall of 3000 to 5000 metres over a distance of 10 to 30 km.

Factor Responsible for Avalanches

There is nobody purpose for the improvement of avalanches. It was accepted for long that the resound of a human voice in the mountains could remove enough snow to begin one. Additionally, a man's weight can cause a avalanches as well. The sudden expansion of weight can break a feeble avalanches. In any case, logical comprehension of torrential slides demonstrates to us that there are numerous ecological elements at work.

1. **Snowstorm and wind direction:** Heavy snowstorms will probably cause avalanches. The 24 hours after a tempest are thought to be the most basic. Twist typically blows from one side of the slant of mountain to another side. While exploding, it will scour snow off the surface which can overhang a mountain.
2. **Substantial snowfall:** Heavy snowfall is the initially, since it stores snow in precarious zones and puts weight on the snow-pack. Precipitation amid the late spring months is the main source of wet snow torrential slides.
3. **Human activity:** Humans have added to the beginning of numerous torrential slides

lately. Winter brandishes that require soak inclines frequently put weight on the snow-pack which it cannot bargain. Joined with the overwhelming deforestation and soil disintegration in mountain areas, it gives the snow little security in the winter months. Promote characteristic causes incorporate seismic tremors and tremors, since they can regularly make splits in the snow-pack.

4. **Vibration or movement:** The utilisation of all terrain vehicles and snowmobiles makes vibrations inside the snow that it cannot withstand. Combined with the gravitational draw, it is one of the speediest approaches to cause a torrential slide. The other is development work finished with explosives, which have a tendency to debilitate the whole encompassing region.

5. **Layers of snow:** There are conditions where snow is as of now on the mountains and has transformed into ice. At that point, new snowfalls on top which can without much of a stretch slide down.

6. **Soak slopes:** Layers of snow develop and slide down the mountain at a quicker rate as steep inclines can expand the speed of snow. A stone or bit of gigantic ice can shake the snow and make it descend.

7. **Warm temperature:** Warm temperatures that can most recent a few hours daily can debilitate a portion of the upper layers of snow and make it slide down.

Impacts of Avalanches

All things considered, there is little harm to the general natural framework because of torrential slides. They are a piece of nature and have been occurring for a large number of years. In any case, they are a noteworthy common danger for the neighbourhood human populace.

1. **Harm to life and property:** An extensive number of losses happens after torrential slides hit intensely populated zones. Framework is harmed and the blockage caused, impacts the work of many. Individuals who appreciate skiing, snowboarding and snowmobiling are at a more serious danger of losing their lives. An effective torrential slide can even annihilate structures and power supplies can be cut off.

2. **Flood:** When a torrential slide happens, it cuts down all the flotsam and jetsam with it and can cause destruction in low lying zones. Streak surges are believed to occur after torrential slides, which are a long haul issue numerous villagers and townspeople need to manage. They can likewise change climate examples and cause edit disappointment in ranches exhibit on the lower fields.

3. **Financial impact:** A torrential slide can piece anything in its way and even limit the typical development of activity. Different ski resorts rely upon vacationers to maintain their business effectively. Ski resorts and different organisations are compelled to close until the point when the torrential slide reductions and climate conditions end up appropriate.

Control Measures

Torrential slide control structures:

Two noteworthy sorts of torrential slide control structures are as prevention structures and prediction structures.

a. **Prevention structures:** These structures are implied for keeping the event of torrential slides. Following are the significant preventive structures.

   i. *Avalanches prevention forest:* These keep the development of torrential slides by the protection of tree trunks and branches, increment the soundness of snow cover by consistently dispersing it and control speedy changes in snow cover.

   ii. *Stepped terraces:* Propositions help in balancing out the snow cover. Ventured porches are anything but difficult to build yet are not successful in controlling surface layer torrential slides.

   iii. *Avalanche control piles:* Torrential slide control piles are gatherings of single heaps crashed into slants in torrential slide zones to control surface layer torrential slides. Dispersing of heaps relies on the sort of snow or land highlights. The normal dispersing is around 5 metres.

iv. *Avalanche control fence:* Torrential slide control fence is introduced on slants of torrential slide zones to anticipate full profundity or surface layer torrential slides.

v. *Suspended fences:* Suspended walls are utilised as a part of soak slants or in regions where establishments cannot be legitimately introduced as a result of poor ground conditions. These are valuable in little zone.

vi. *Snow cornice control structures:* These structures are introduced at highest points of mountain regions to keep the improvement of snow cornices that can cause torrential slides.

b. **Protection structures:** These structures are introduced in the way of the torrential slide or in snow store zones to alter the stream of course of torrential slides, to decrease their vitality to hinder their stream or to permit their section. Following are the primary defensive structures.

i. *Protective fences:* These are introduced to hinder the torrential slides and their activity is like that of holding dividers. They are regularly developed of steel and are utilized fundamentally to block little torrential slides.

ii. *Retaining walls:* Holding dividers are regularly introduced in snow store territories to hinder the stream of torrential slides previously they achieve the roadside. These dividers require a pocket sufficiently extensive to store snow kept by torrential slides and are not extremely successful unless they are introduced on delicate inclines of 20° or less.

iii. *Deflecting structures:* As the name demonstrates, these structures are introduced to divert the stream of a torrential slide. This is done especially to maintain a strategic distance from obstruction of the torrential slide in street movement.

iv. *Snow sheds:* Snow shed is a roofed structure introduced over a street to permit the stream of a torrential slide over the rooftop. This is most solid of the different torrential slide assurance structures.

v. *Retarding structures:* These are structures to decrease the stream speed or the size of the torrential slides. There are different writes, for example, earth hills, impeding heaps, grinding bunk work and hindering wall.

Other Control Measures

Aside from the previously mentioned measures, there are other control measures which are quickly depicted as under.

i. **Prediction and forecasting:** Expectation and determining is an extremely successful strategy for lessening the hazard from torrential slides can forestall torrential slide debacles as well as make it effectively discard unsafe snow stores and cornices.

ii. **Disposal of avalanches potential snow packs:** Techniques that discard snow packs on risky slants incorporate hard work, mechanical strategies that utilisation impacting powder. All in all, little torrential slides are generally discarded by impacting.

HEAT WAVE

A heat wave is a period of excessively hot weather, which may be accompanied by high humidity, especially in oceanic climate countries. Extreme positive departures from the normal maximum temperature result in a heat wave during the summer season. The rising maximum temperature during the pre-monsoon months often continues till June, in rare cases till July, over the northwestern parts of the country.

Lately, heat wave actuated losses have somewhat expanded. Strangely high temperatures were seen amid April 2002 the nation over. On tenth May 2002, the greatest temperature at Gannavaram (Vijayawada) 49°C was recorded. Mortality and uneasiness expanded because of least temperatures in summer do not permit the important night-time cooling to kill the high greatest temperature amid a heat wave age. Heat waves in India in the diurnal temperature range (DTR) because of urbanisation is another factor prompting human.

COLD WAVES

A cold wave (referred to in a few districts as a chilly spell or icy spell) is a climate change that is recognised by a cooling of the air. In particular, as utilised by the US National Weather Service, a cool wave is a quick fall in temperature inside a 24-hour time frame requiring generously expanded assurance to farming, industry, business, and social exercises. The exact basis for an icy wave is controlled by the rate at which the temperature falls, and the base to which it falls. This base temperature is reliant on the topographical district and time of year.

Occurrence of extreme low temperature in relationship with attack of dry cool twists from north into the sub landmass are known as cold waves. The Northern parts of India, extraordinarily the bumpy locales and the connecting fields are impacted by transient unsettling influences in the mid-scope westerlies which regularly have feeble frontal qualities. These are known as Western unsettling influences. The cold waves chiefly influence the territories toward the north of 20°N yet in relationship with huge plentifulness troughs, cool wave conditions are some of the time announced from states like Maharashtra and Karnataka also. As of late because of decay of the air quality in urban areas of India the passing and inconvenience from cool waves have been generous. UP and Bihar rank the most noteworthy as far as setbacks from cool wave and this could be because of poor level of improvement and absence of safe houses to the open air specialists and ranchers.

VOLCANIC ERUPTIONS

Volcanoes can cause widespread destruction and consequent disaster in several ways. The effects include the volcanic eruption itself that may cause harm following the explosion of the volcano or falling rocks. Secondly, lava may be produced during the eruption of a volcano, and so as it leaves the volcano the lava destroys many buildings, plants and animals due to its extreme heat. Thirdly, volcanic ash, generally meaning the cooled ash, may form a cloud, and settle thickly in nearby locations. When mixed with water this forms a concrete-like material. In sufficient quantities, ash may cause roofs to collapse under its weight but even small quantities will harm humans if inhaled. Since the ash has the consistency of ground glass it causes abrasion damage to moving parts such as engines. The main killer of humans in the immediate surroundings of a volcanic eruption is the pyroclastic flows, which consist of a cloud of hot volcanic ash which builds up in the air above the volcano and rushes down the slopes when the eruption no longer supports the lifting of the gases. It is believed that Pompeii was destroyed by a pyroclastic flow. A lahar is a volcanic mudflow or landslide. The 1953 Tangiwai disaster was caused by a lahar, as was the 1985 Armero tragedy in which the town of Armero was buried and an estimated 23,000 people were killed.

A specific type of volcano is the super volcano. According to the Toba catastrophe theory, 75,000 to 80,000 years ago a super volcanic event at Lake Toba reduced the human population to 10,000 or even 1,000 breeding pairs, creating a bottleneck in human evolution. It also killed three-quarters of all plant life in the Northern hemisphere. The main danger from a super volcano is the immense cloud of ash, which has a disastrous global effect on climate and temperature for many years.

Volcanic emissions and seismic tremors are a route for earth to discharge weight and warmth, much like a well-being valve. There are three commanding hypotheses to clarify what makes a well of lava eject.

Thickness Distinction in Magma

According to the primary hypothesis, because of warmth and weight in the World's mantle, strong rocks dissolve, to frame magma. Magma has an indistinguishable mass from the strong shake, yet more volume, influencing it to lighter and more light. Along these lines, it will endeavour to rise, if this magma keeps on experiencing high-thickness material till it achieves the World's outside layer, volcanic ejection happens. It can either be as a magma stream or might be hazardous.

Weight of Discharged Gases

As per the second hypothesis, magma contains broke up substances, for example, water, sulphur dioxide and carbon dioxide. The dissolvability of magma diminishes with the reduction in weight as it ascends towards the outside layer, and the gases get discharged as air pockets. At the point when the volume of the gas rises in magma stretches around 75%, magma deteriorates into pyroclasts, a blend of in part liquid and strong sections. The blasting of pyroclasts is extremely dangerous and the reason for probably the most fierce ejections on the surface of earth.

Infusion of New Magma

The third hypothesis says that when new magma enters a chamber officially overflowing with magma, the spring of gushing lava emits because of the extra weight applied by the infusion of new magma. This kind of emission can be tranquil or brutal.

The power of ejection as a rule relies upon the consistency of magma and its gas content. High-thickness magma more often than not brings about greater, more exceptional emissions, though magma that streams effortlessly will have bring down weight develop, so a less brutal ejection. Profoundly thick magma is described by the nearness of more silicates and contains less broke down water. Another critical factor is the measure of gases exhibit in the magma. Magma containing huge measure of caught gases will prompt a fierce ejection, and less gases in magma will bring about an unrestrained stream.

Types of Volcanic Eruption

Contingent upon their force, volcanic ejections can be separated into the accompanying real composes.

Regular to the islands of Hawaii, these ejections are normally connected with hotspots, and have non-unstable magma streaming out of gaps or vents, frequently on the slants. At the point when the focal vent ejects, the spring of gushing lava hurls a fire wellspring splendid orange magma splashing into the air for a few hours or couple of minutes. The magma has almost no gas substance and creeps gradually to frame a magma lake.

Strombolian Eruptions

Pressing more power than the Hawaiian, Strombolian ejections are described by short hazardous blasts joined by blasting sounds. Caused by blasting gas bubbles, the emissions can shape segments up to 100 meters tall and can most recent a few centuries.

Vulcanian Eruptions

Vulcanian ejections work in a comparative way to the strombolian emissions, aside from that these are substantially more grounded and subsequently more ruinous, however with time, their arch gets harmed and prompts a more nonstop magma stream rather than short blasts of magma.

Plinian Eruptions

The most intense of all emissions, they have caused enormous harm, spreading a large number of miles, pulverising urban areas and changing the atmosphere. These emissions are caused by extremely thick magma with a high gas content, shaping tall segments of gas, and cinder, looking like mushroom mists, from an atomic blast, more than 35 miles high and can keep going for a considerable length of time. The tephra, particularly when joined with softened snow, streams amazingly quick and singes everything in its way. These emissions regularly happen suddenly, staying inert for quite a long-time, getting life frames unprepared as they scramble to get away from its pyroclastic stream, harmful gases and choking out fiery remains mists.

Pelean Eruptions

These are fundamentally the same as the Plinian emissions and are similarly ruinous. The Pelean emission's significant harm originates from torrential slides and avalanches of gleaming fiery remains streaming down the precarious slants at monstrous paces though Plinian eruption has tall sections of powder and smoke.

In spite of the fact that geologists have clarified a considerable lot of the secrets, more up-to-date revelations ceaselessly challenge their hypotheses. Volcanologist persistently work hard trying to better comprehend what makes a spring of gushing lava eject. By understanding this marvel, we may have the capacity to decrease its effect on human life, and even outfit its energy to create power.

CLIMATE CHANGE

Climate change is any long-term significant change in the "average weather" of a region or the earth as a whole. Average weather may include average temperature, precipitation and wind patterns. It involves changes in the variability or average state of the atmosphere over durations ranging from decades to millions of years. These changes can be caused by dynamic processes on earth, external forces including variations in sunlight intensity, and more recently by human activities. Weather is the day-to-day state of the atmosphere, and is a chaotic non-linear dynamical system. On the other hand,

climate, the average state of weather, is fairly stable and predictable. Climate includes the average temperature, amount of precipitation, days of sunlight, and other variables that might be measured at any given site. However, there are also changes within the earth's environment that can affect the climate. In recent usage, especially in the context of environmental policy, the term "climate change" usually refers to changes in modern climate like 'global warming'. Climate changes reflect variations within the earth's atmosphere, processes in other parts of the earth such as oceans and ice caps, and the effects of human activity. The external factors that can shape climate include such processes as variations in solar radiation, the earth's orbit, and greenhouse gas concentrations. Current studies indicate that radioactive forcing by greenhouse gases is the primary cause of global warming. Greenhouse gases are also important in understanding earth's climate history. The greenhouse effect, which is the warming produced as greenhouse gases trap heat, plays a key role in regulating earth's temperature.

Most atmosphere researchers concur the fundamental driver of the momentum an unnatural weather change drift is human extension of the "nursery effect" — warming that outcomes when the environment traps warm emanating from earth toward space.

Certain gases in the environment piece warm from getting away. Seemingly perpetual gases that remain same for all time in the air and do not react physically or synthetically to changes in temperature are depicted as "driving" environmental change. Gases, for example, water vapour, which react physically or artificially to changes in temperature, are viewed as "criticisms."

Gases that contribute to the greenhouse effect have been included in Fig. 1.4.

Water Vapour

The most plentiful ozone harming substance, however significantly, it goes about as a criticism to the atmosphere. Water vapour

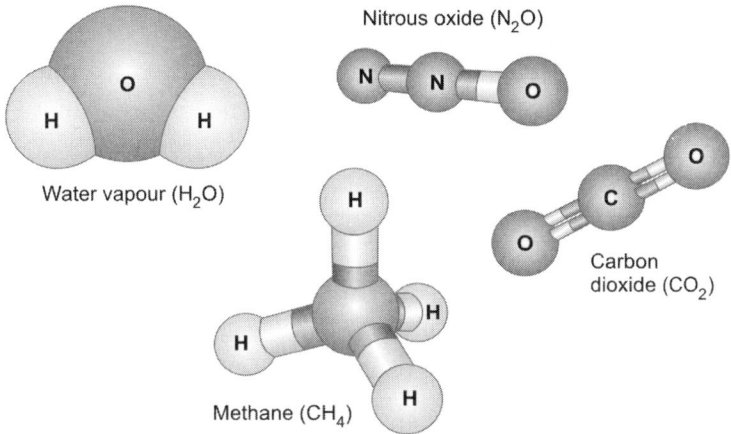

Fig. 1.4: Gases responsible for GHG emission

increments as the earth's air warms, yet so does the likelihood of mists and precipitation, making these the absolute most imperative criticism instruments to the nursery impact.

Carbon Dioxide (CO_2)

A minor however essential segment of the environment, carbon dioxide is discharged through normal procedures, for example, breath and fountain of liquid magma emissions and through human exercises, for example, deforestation, arrive utilize changes, and copying petroleum products. People have expanded climatic CO_2 focus by in excess of a third since the Industrial Revolution started. This is the most vital enduring "compelling" of environmental change.

Methane

A hydrocarbon gas delivered both through common sources and human exercises, including the decay of squanders in landfills, agribusiness, and particularly rice development, and additionally ruminant absorption and fertilizer administration related with local domesticated animals. On an atom for particle premise, methane is a significantly more dynamic ozone depleting substance than carbon dioxide, yet in addition one which is considerably less rich in the environment.

Nitrous Oxide

A capable ozone depleting substance created by soil development hones, particularly the utilisation of business and natural manures, non-renewable energy source ignition, nitric corrosive generation, and biomass consuming.

Chlorofluorocarbons (CFCs)

Manufactured mixes completely of mechanical beginning utilised as a part of various applications, however now to a great extent directed underway and discharge to the climate by worldwide understanding for their capacity to add to devastation of the ozone layer. They are additionally ozone depleting substances.

On earth, human exercises are changing the common nursery. In the course of the most recent century the consuming of petroleum derivatives like coal and oil has expanded the centralization of barometrical carbon dioxide (CO_2). This happens in light of the fact that the coal or oil consuming procedure consolidates carbon with oxygen noticeable all around to make CO_2. To a lesser degree, the clearing of land for farming, industry, and other human exercises has expanded groupings of ozone depleting substances.

The outcomes of changing the characteristic air nursery are hard to anticipate, however certain impacts appear to be likely:

All things considered, earth will end up hotter. A few districts may welcome hotter temperatures, yet others may not.

Hotter conditions will most likely prompt more dissipation and precipitation generally, yet singular locales will fluctuate, some getting to be wetter and others dryer.

A more grounded nursery impact will warm the seas and halfway dissolve icy masses and other ice, expanding ocean level. Seawater likewise will grow on the off chance that it warms, contributing further to ocean level ascent.

In the interim, a few yields and different plants may react positively to expanded air CO_2, developing all the more overwhelmingly and utilising water all the more effectively. In the meantime, higher temperatures and moving atmosphere examples may change the zones where crops develop best and influence the cosmetics of common plant groups.

GLOBAL WARMING

Global warming occurs when carbon dioxide (CO_2) and other air pollutants and greenhouse gases collect in the atmosphere and absorb sunlight and solar radiation that have bounced off the earth's surface. Normally, this radiation would escape into space—but these pollutants, which can last for years to centuries in the atmosphere, trap the heat and cause the planet to get hotter. That's what is known as the greenhouse effect.

Global warming is the term used to portray a progressive increment in the normal temperature of the earth's air and its seas, a change that is accepted to be for all time changing the earth's atmosphere. There is awesome verbal confrontation among numerous individuals, and now and again in the news, on whether a dangerous atmospheric devotion is genuine (some call it a fabrication). However, atmosphere researchers taking a gander at the information and certainties concur the planet is warming. While numerous view the impacts of an earth-wide

temperature boost to be more significant and more quickly happening than others do, the logical accord on climatic changes identified with an earth-wide temperature boost is that the normal temperature of the earth has ascended in the vicinity of 0.4° and 0.8°C in the course of recent years. The expanded volumes of carbon dioxide and other ozone harming substances discharged by the copying of petroleum products, arrive clearing, horticulture, and other human exercises are accepted to be the essential wellsprings of the worldwide temperature alteration that has happened in the course of recent years. Researchers from the intergovernmental panel on climate completing an earth-wide temperature boost investigate have as of late anticipated that normal worldwide temperatures could increment in the vicinity of 1.4° and 5.8°C by the year 2100. Changes coming about because of an earth-wide temperature boost may incorporate rising ocean levels because of the softening of the polar ice tops, and in addition an expansion in event and seriousness of tempests and other extreme climate occasions.

Global warming, also referred to as climate change, is the observed century-scale rise in the average temperature of the earth's climate system and its related effects (Gillis, 2015). Multiple lines of scientific evidence show that the climate system is warming (Hartmann et al, 2013). Many of the observed changes since the 1950s are unprecedented in the instrumental temperature record which extends back to the mid-19th century, and in paleoclimate proxy records covering thousands of years (IPCC, 2013).

Causes

On earth, human exercises are changing the common nursery. In the course of the most recent century the consuming of petroleum derivatives like coal and oil has expanded the centralisation of barometrical carbon dioxide (CO_2). This happens in light of the fact that the coal or oil consuming procedure consolidates carbon with oxygen noticeable all around to make CO_2. To a lesser degree, the clearing of land for farming, industry, and other human exercises has expanded groupings of ozone depleting substances.

The outcomes of changing the characteristic air nursery are hard to anticipate, however, certain impacts appear to be likely:

All things considered, earth will end up hotter. A few districts may welcome hotter temperatures, yet others may not.

Hotter conditions will most likely prompt more dissipation and precipitation generally, yet singular locales will fluctuate, some getting to be wetter and others dryer.

A more grounded nursery impact will warm the seas and halfway dissolve icy masses and other ice, expanding ocean level. Seawater likewise will grow on the off chance that it warms, contributing further to ocean level ascent.

In the interim, a few yields and different plants may react positively to expanded air CO_2, developing all the more overwhelmingly and utilising water all the more effectively. In the meantime, higher temperatures and moving atmosphere examples may change the zones where crops develop best and influence the cosmetics of common plant groups.

Impacts of an Unnatural Weather Change

Each year, researchers take in more about the results of a worldwide temperature alteration, and numerous concur that ecological, monetary, and well-being outcomes are probably going to happen if flow patterns proceed. Here is only a sprinkling of what we can anticipate

- Liquefying icy masses, early snowmelt, and extreme dry spells will cause more emotional water deficiencies and increment the danger of out of control fires in the American West.
- Rising ocean levels will prompt seaside flooding on the Eastern Seaboard, particularly in Florida, and in different regions, for example, the Gulf of Mexico.

- Woods, homesteads, and urban areas will confront troublesome new bugs, warm waves, overwhelming deluges, and expanded flooding. Each one of those components will harm or devastate horticulture and fisheries.
- Interruption of living spaces, for example, coral reefs and Alpine knolls could drive numerous plant and creature species to annihilation.
- Sensitivities, asthma, and irresistible ailment flare-ups will turn out to be more typical because of expanded development of dust creating ragweed, larger amounts of air contamination, and the spread of conditions good to pathogens and mosquitoes.

Preventive Measures

1. Buy a fuel-proficient auto (evaluated at 32 mpg or more) to supplant your most habitually utilised car.
2. Protect your home, clean your ventilating channels and introduce vitality effective showerheads.
3. Leave your auto at home (walk, bicycle or take mass travel).
4. Reuse your home's waste newsprint, cardboard, glass and metal.
5. Introduce a sunlight-based warmed framework to give your high temperature water.
6. Supplant brilliant lights with reduced bright light bulbs.
7. Purchase sustenance and different items with reusable or recyclable bundling rather than those in non-recyclable bundling.
8. Supplant your present home apparatuses (cooler, clothes washer, dish washer) with high-productivity models.
9. Utilise an electric or drive cutter rather than a fuel controlled trimmer to cut your grass.
10. Plant local, dry season safe trees and bushes around your home and open air aerating and cooling unit.

SEA LEVEL RISE

An ocean level ascent is an expansion in worldwide mean ocean level because of an increment in the volume of water on the planet's seas. Ocean level ascent is generally credited to worldwide environmental change by warm extension of the water in the seas and by softening of ice sheets and icy masses on land (Shennan, 2013). The liquefying of drifting ice racks and ice shelves adrift would raise ocean levels just by around 4 cm (1.6 inch) (Noerdlinger et al, 2007).

Ocean level ascent at particular areas might be pretty much than the worldwide normal. Nearby factors may incorporate structural impacts, subsidence of the land, tides, streams, storms, etc. (Fischlin et al, 2007). Sea level ascent is relied upon to proceed for quite a long-time. As a result of long reaction times for parts of the atmosphere framework, it has been evaluated that we are as of now dedicated to an ocean level ascent inside the following 2,000 years of roughly 2.3 metres (7.5 ft) for every degree Celsius of temperature rise. (Levermann et al, 2013). The international Panel on Climate Change (IPCC) Summary for Policymakers, 2014, predicts that the worldwide mean ocean level ascent will keep amid the 21st century, likely at a speedier rate than saw from 1971 to 2010. Projected rates and sums fluctuate. A January 2017 NOAA (National Oceanic and Atmospheric Administration) report recommends a scope of GMSL (Global Mean Sea Level) ascent of 0.3 – 2.5 m conceivable amid the 21st century.

Two primary causes to sea level rise: (i) Warm development of ocean water because of sea warming and (ii) water mass contribution from arrive ice dissolve and land water supplies. Sea temperature information gathered amid the previous couple of decades show that sea warm development has altogether expanded amid the second 50% of the twentieth century. Warm development represents around 25% of the watched SLR since 1960 and around half from 1993 to 2003. From that point forward, upper-sea warming

has been littler and by and large finished the satellite altimetry period (1993 to 2009), the commitment of sea temperature change to the worldwide mean ocean level might be ~30%.

Since the mid 1990s, diverse remote-detecting devices [airborne and satellite radar and laser altimetry; manufactured gap radar interferometer (InSAR); and, since 2002, space gravimetry from the gravity recovery and climate experiment (GRACE) mission have given great information on the mass adjust of the polar ice sheets. These pieces of information show that Greenland and West Antarctica mass misfortune is quickening. In the vicinity of 1993 and 2003, <15% of the worldwide SLR was because of the ice sheets. In any case, since around 2003, their commitment has about multiplied; expanding icy mass and ice sheet mass misfortune has made up for decreased sea warm extension, with the end goal that SLR proceeds at nearly a similar rate. In spite of the fact that not monotonic through time, we assess that by and large finished the altimetry time (1993 to 2009), add up to arrive ice mass misfortune clarifies ~60% of the rate of SLR.

Quickened loss of ice sheet mass somewhat comes about because of fast outlet ice sheet stream along a few edges of Greenland and West Antarctica where the establishing line is underneath ocean level, and further ice shelf release into the encompassing sea. Late perceptions recommend that warming of subsurface sea waters triggers beach front ice release. Albeit surface mass procedures (snow gathering versus surface dissolving) likewise add to Greenland mass misfortune, in West Antarctica mass misfortune basically comes about because of ice flow.

Satellite altimetry demonstrates that ocean level is not rising consistently. In a few areas (e.g. Western Pacific), ocean level has ascended to three times quicker than the worldwide mean since 1993. Spatial examples in ocean level patterns mostly result from non-uniform sea warming and saltiness varieties, albeit different factors likewise contribute, including the strong earth reaction to the last deglaciation and gravitational impacts and changes in sea dissemination because of continuous land ice liquefying and freshwater input. Spatial examples in sea warm extension are not lasting highlights: They vary in space and time in light of regular bothers of the atmosphere framework; thus, we expect that the ocean level change examples will waver on multi-decadal time scales. IPCC projections propose apparent provincial changeability around the future worldwide mean ascent by 2100 in light of non-uniform future sea warming, however, understanding between the models is poor. Be that as it may, precise evaluations of future provincial ocean level changes are required for beach front effect and adjustment appraisal.

Impacts of Sea Level Rise

The impact of sea level rise is submergence and expanded flooding of seaside arrive, and in addition saltwater interruption of surface waters. Longer-term impacts additionally happen as the drift changes with the new conditions, including expanded disintegration and saltwater interruption into groundwater. Beach front wetlands, for example, salt marshes and mangroves will likewise decrease unless they have an adequate dregs supply to keep pace with SLR. These physical effects thus have both immediate and roundabout financial effects, which give off an impression of being overwhelmingly negative. Despite the fact that atmosphere initiated SLR is imperative, waterfront impacts additionally result from relative (or neighbourhood) SLR (e.g. from geographical procedures, for example, subsidence). For instance, relative ocean level is by and by falling where arrive is inspiring impressively, for example, the Northern Baltic and Hudson Bay—the destinations of substantial (kilometre-thick) ice sheets amid the last frigid greatest. Conversely, relative ocean level is rising more quickly than atmosphere instigated drifts on dying down coasts. In numerous areas, human exercises are fueling subsidence on defenseless drifts, including most stream deltas [e.g. the Ganges—Brahmaputra, Mekong, and Changjiang

deltas]. The most sensational subsidence impacts have been caused by waste and groundwater liquid withdrawal; over the twentieth century, coasts have died down by up to 5 m in Tokyo, 3 m in Shanghai, and 2 m in Bangkok. To maintain a strategic distance from submergence and additionally visit flooding, these urban areas now all rely upon a generous surge safeguard and water administration foundation. South of Bangkok, subsidence has prompted considerable shoreline withdrawn of in excess of 1 km, leaving utility poles remaining in the ocean.

As the greatness of atmosphere incited SLR expands, the effects will turn out to be more evident, particularly in certain low-height waterfront zones. Most nations in South, Southeast, and East Asia have all the earmarks of being very undermined due to across the board event of thickly populated deltas, regularly connected with expansive developing urban areas. Africa likewise shows up profoundly debilitated inferable from the low levels of advancement joined with desires of fast populace development in beach front zone: Egypt and Mozambique are two "hotspots" for potential effects. Be that as it may, the little island states encounter the biggest relative increment in impacts, including areas of high islands like the Caribbean. Low islands, for example, the Maldives or Tuvalu confront the genuine prospect of submergence and finish relinquishment amid the 21st century.

OZONE DEPLETION

Ozone exhaustion alludes two related wonders saw. One is a persisting decay of around 4% in the aggregate total of ozone in earth's stratosphere (the ozone layer), and a through and through more prominent springtime diminish in stratospheric ozone around earth's polar regions. The last contemplate is suggested as the ozone opening. There are in like way springtime polar tropospheric ozone use occasions regardless of these stratospheric ponders.

The main causes of ozone depletion is man-made chemicals, especially man-made halocarbon refrigerants, solvents, powers, and foam blowing administrators (chlorofluorocarbons (CFCs), HCFCs, halons), implied as ozone-depleting substances (ODS). ODS discharge halogen iotas in the stratosphere through photo dissociation, which catalyse the breakdown of ozone (O_3) into oxygen (O_2). Both kinds of ozone exhaustion were seen to increment as outflows of halocarbons expanded.

Ozone exhaustion and the ozone gap have produced overall worry over expanded malignancy dangers and other negative impacts. The ozone layer balances most pernicious UVB wavelengths of splendid light (UV light) from experiencing the earth's condition. These wavelengths cause skin growth, sunburn, and waterfalls, which were anticipated to increment significantly because of diminishing ozone, and additionally hurting plants and creatures.

Ozone utilisation, non-stop lessening of earth's ozone layer in the upper condition caused by the landing of engineered blends containing vaporous chlorine or bromine from industry and other human activities. The diminishing is most articulated in the polar locales, particularly finished Antarctica. Ozone consumption is a noteworthy ecological issue since it expands the measure of bright (UV) radiation that achieves earth's surface, which builds the rate of skin growth, eye waterfalls, and hereditary and insusceptible framework harm. The Montreal Protocol, confirmed in 1987, was the first of a few far reaching global attentions authorised to end the generation and utilisation of ozone-exhausting chemicals. Because of proceeded with worldwide collaboration on this issue, the ozone layer is required to recoup after some time.

Questions

1. Explain the meaning of the words natural 'Hazard' and 'Disaster'.
2. What are various types of environmental hazards? Differentiate between hazard and disaster
3. What is capacity and risk?
4. Recognise and describe some disaster-prone areas from each physical division of India.
5. Describe some adverse effects of natural disasters.
6. Suggest measures to mitigate or reduce the problems and sufferings arising before, during or after the disaster.
7. Name any two causes of flood.
8. How much area of the country is flood prone?
9. Name any two measures of flood control.
10. What monitoring systems are used for tracing the path of cyclones?
11. What is a drought?
12. What are major causes of draught?
13. List three pre-disaster activities to reduce the impact of cyclones.
14. How heat waves formed?
15. Name any two causes of landslide.
16. What are the mitigation measures of landslides?
17. Write some do's and don'ts during and after the earthquake.
18. What is an earthquake? List out the causes of an earthquake.
19. Identify three major mitigation measures to reduce earthquake risk.
20. What name is given to the high sea-wave triggered by earthquake?
21. Which are the most cyclone prone months?
22. What is a landslide? What activities of human being have increased the frequency of landslides?
23. When do cyclones occur in India? Describe the measures adopted for protection from the cyclones.
24. What are the effects of cold waves?
25. What is sea level rise and how it formed?
26. How ozone depletion impacted on environment?

2

Man-Made Disaster

Disaster can take a wide range of structures, and the duration can go from a hourly disturbance to days or long stretches of continuous decimation. The following is a rundown of the different sorts of disasters—both normal and man-made or innovative in nature—that can affect a network. Calamities additionally can be caused by people. Perilous materials crises incorporate concoction spills and groundwater defilement. Working environment fires are more typical and can cause critical property harm and death toll. Networks are additionally helpless against dangers postured by fanatic gatherings who utilise viciousness against the two individuals and property.

High-hazard targets incorporate military and non-military personnel government offices, universal air terminals, substantial urban communities and prominent points of interest. Digital fear mongering includes assaults against PCs and systems done to threaten or force an administration or its kin for political or social targets.

Our condition has been the casualty of a wide range of assaults. A portion of these assaults are common, for example, tropical storms and quakes. Notwithstanding, there are assaults that are unnatural and man-made, for example, wars, blasts, concoction spills, and so forth. These assaults for the most part convey with them overwhelming sticker prices as property and lives are harmed past full remuneration and repair.

On the basis of their origin, disasters are categorised as natural or man-made. How-ever, there may be no clear cut boundary between them. The High Power Committee on Disaster Management (HPCDM) of the Govt. of India constitute in 1999, identified 31 various disasters and categorised them into five major categorises, primarily based on their origin.

These are:

Sub-group-1: Water and climate related hazards

Sub-group 2: Geologically related hazards

Sub-group 3: Chemical, industrial and nuclear related disasters

Sub-group 4: Accident related disasters

Sub-group 5: Biologically related disasters

Disaster can also be classified by nature, timing, predictability, response time and type of impact (*see* Chapter 1).

NUCLEAR DISASTER

International Atomic Energy Agency defines Nuclear Disaster as "an occasion that has prompted noteworthy results to the general population, the earth or the office." For example, incorporate deadly impacts to people, extensive radioactivity discharge into nature, or reactor center soften.

A nuclear disaster could take several forms. The most evident would be an emergency at an atomic reactor plant. Though the plant might not explode, the result of such a disaster would very likely the release of massive amount of radiation and radioactive material into the environment and it would take hundred years to rot to anything close "safe" levels.

Specialised measures should be adopted to reduce the ill effect of disasters or to limit the measure of radioactivity discharged to the earth.

There are many forms of nuclear disaster. The most evident would be an emergency at a nuclear reactor plant. In spite of the fact that the plant will not detonate, the aftereffect of such a calamity would likely the arrival of gigantic measure of radiation and radioactive material into nature and it would take hundred years to rot to anything close "safe" levels.

Causes of Nuclear Disaster

Nuclear disasters are generally connected with meltdowns. At the point when an emergency happens in a reactor, the reactor "liquefies". That is, the temperature ascends in the center so much that the fuel poles really swing to fluid, similar to ice transforms into water when warmed. On the off chance that the centre keeps on warming, the reactor would get so hot that the steel dividers of the centre would likewise dissolve. In an entire reactor emergency to a great degree hot (around 2700° Celsius) liquid uranium fuel poles would dissolve through the base of the reactor and really sink around 50 feet into the earth underneath the power plant.

The liquid uranium would respond with groundwater, creating expansive blasts of radioactive steam and flotsam and jetsam that would influence close-by towns and populace focuses.

Consequences of Nuclear Disasters

Nuclear blasts create both quick and damaging impacts. Quick impacts are delivered and cause critical pulverization inside seconds or minutes of an atomic explosion. The deferred impacts (radioactive aftermath and other conceivable condition impacts) deliver harm over a stretched out period going from hours to hundreds of years, and can cause antagonistic impacts in areas extremely removed from the site of the explosion.

Nuclear disaster can create atmosphere issues in light of the fact that the high temperatures of the atomic fireball make a lot of nitrogen oxides frame from the oxygen and nitrogen in the environment (fundamentally the same as what occurs in burning motors). Every megaton of yield will create exactly 5000 tonnes of nitrogen oxides. The rising fireball of a high kiloton or megaton extend warhead will convey these nitric oxides spring up into the stratosphere, where they can achieve the ozone layer. A progression of vast barometrical blasts could fundamentally exhaust the ozone layer.

Mitigation Strategies

Measures are insufficient to totally mitigate the destructive impacts of a nuclear disaster. A nuclear debacle causes the depletion of the ozone layer which thusly prompts skin diseases. The best way to guarantee well-being of individuals is to manufacture powerful atomic reactors and proficient coolants. Anticipation is superior to fix.

The mitigation measures could be as far as decrease of death toll, harm to property, and consequential impact on future generations, radiation consequences for different natural types of life—on people, creatures, and plants. Any activity to anticipate utilisation of atomic weapons in wars could likewise be a part of the mitigation measures.

Suitable strategic plan, installation, activity are done in atomic reactors, there is a plausibility of little missteps. An atomic reactor is the most advanced method for warming water to extremely hot steam. It utilises the warmth vitality from an atomic splitting chain response. A few times obscure and unforeseen episodes happen like in Fukushima in Japan in 2011. Some dread additionally exists because of psychological warfare, if fear based oppressors undermine a reactor or cause catastrophe. According to record there has been a noteworthy mischance in every decade over the most recent 5 decades. There have been around 100 atomic mishaps little or enormous up until this point.

So, there any way to forestall atomic debacles to happen, or possibly—on the off chance that such an occasion cannot be evaded – what should be possible to decrease however much as could be expected serious results for men and condition? The most effortless path, as some claim, could be to evade any further utilisation of atomic innovation. This is to state, no atomic power plants, atomic weapons, or some other utilisations of atomic material and ionising radiation in inquire about, innovation, solution, and so forth. This, be that as it may, is not extremely practical. No moral exchanges will stop logical and mechanical advance. They could or should, despite what might be expected, add to enhance well-being and security of such specialised uses of atomic science and not just preclude their further utilise. In the field of atomic weapons, there has been some advance. Toward the finish of the Apartheid administration, South Africa ceased their atomic weapons program and destroyed (under close observation by specialists of the IAEA) their atomic warheads making the Southern half of the globe is currently an atomic weapon-free zone. This is authorized by the Pelindaba Treaty, which entered in compel on July fifteenth 2009. Also, the Treaty on the non-proliferation of nuclear weapons (NPT, non-proliferation arrangement for atomic weapons), that entered in compel 1970 and was reached out in 1995, gives some extra security for mankind.

CHEMICAL DISASTERS

Chemical, being at the centre of present day modern frameworks, has accomplished an intense pressure for disaster management government, NGOs and community.

Chemical disasters might be horrendous in their effects on people and have brought about the setbacks and furthermore harms nature and property. The components which are at most elevated dangers because of chemical disasters fundamentally incorporate the industrial plant, its representatives and specialists, risky synthetics vehicles, the inhabitants of close-by settlements, nearby structures, tenants and encompassing network.

India has seen the world's most noticeably world worst chemical disaster "Bhopal Gas Tragedy" in the year 1984. The Bhopal Gas disaster was most wrecking compound mishap ever, where more than a large number of individuals passed on because of incidental arrival of lethal gas methyl isocyanate (MIC).

Such mishaps are significant in terms of injuries, torment, enduring, loss of lives, harm to property and condition. India kept on seeing a progression of substance mishaps even after Bhopal had shown the defenselessness of the nation. Just in a decade ago, 130 noteworthy concoction mischances announced in India, which came about into 259 passings and 563 number of major harmed.

There are around 1861 major accident hazard (MAH) units, spread crosswise over 301 areas and 25 states and 3 Union Territories, in all zones of nation. Plus, there are a great many enrolled and dangerous manufacturing plants (underneath MAH criteria) and un-sorted out areas managing various scopes of risky material posturing genuine and complex levels of catastrophe dangers. Security activities taken in India to address compound hazard.

Causes of Chemical Disasters

1. Chemical disasters are dangers to individuals and life system that emerge from the large scale manufacturing of products and enterprises. At the point when these dangers surpass human adapting abilities or the absorptive limits

of ecological frameworks they offer ascent to industrial disasters.

2. Modern dangers can happen at any phase in the creation procedure, including extraction, preparing, fabricate, transportation, stockpiling, utilise, and transfer. Misfortunes for the most part include the arrival of harming substances (e.g. synthetic compounds, radioactivity, hereditary materials) or harming levels of vitality from modern offices or hardware into encompassing conditions. This as a rule happens as blasts, fires, spills, holes, or squanders.

3. Losses may happen due to factors that are inner to the modern framework (e.g. designing blemishes) or they may happen in light of outer components (e.g. extremes of nature).

4. The reasons for chemical disasters and debacles are breakdowns, disappointments, or unforeseen symptoms of innovative frameworks. In any case, this is a deceptive misrepresentation and numerous different components are included.

5. The cause of industrial disaster is a mix of mechanical frameworks, individuals, and situations that likewise incorporate topographical, barometrical, biological, mental and social parts.

6. Explosion in a plant dealing with or creating dangerous substances.

7. Accidents away offices taking care of vast and different amounts of synthetic concoctions.

8. Accidents amid the transportation of synthetic substances starting with one site then onto the next.

9. Misuse of synthetic substances, bringing about defilement of nourishment stocks or the earth, overdosing of agrochemicals.

10. Instantaneous mass arrival of poisons/contaminant.

11. Mass discharge from common source.

12. Improper squander administration, for example, uncontrolled dumping of poisonous.

13. Chemicals, disappointment in squander administration frameworks or mischances in wastewater treatment plants.

14. Technological framework disappointments.

15. Failures of plant security outline or plant segments.

16. Natural perils, for example, fire, seismic tremors, avalanches.

17. Sabotage.

18. Mass harming (deliberate or accidental).

19. Human blunder.

20. Manufacturing and formulation facility (counting amid commissioning and process operation; maintenance, disposal and waste management).

21. Material handling and storage.

22. Storage of fuels (LPG depots and so on)
    - Pipelines, and
    - Transportation (street, rail, air, and conduits).

Effects

- Death, damage, physiological well-being impacts and misfortunes.
- Damage to ecological assets, similar to arrive/soil, arrive utilise, water bodies/assets, air-quality and developments, nearby atmosphere, crops/woods and bio-items.
- Disruption of natural administrations, for example, water supply, tasteful and entertainment, ecological and general well-being, sanitation, rubbish administration.
- Damage and misfortunes to structures, structures, machines/hardware, offices.
- Psychological injury, stress and absence of prosperity.
- Insurance misfortunes, and monetary misfortunes identified with disturbance of efficiency, compensation, compensation, motivating forces.
- Increase in weakness to different perils including regular and natural exposures.
- Components deciding seriousness of a risky occasion.

BIOLOGICAL DISASTERS

Biological disasters are phenomenon of natural inception or passed on by natural vectors, including introduction to pathogenic micro-organisms, poisons and bioactive substances that may cause death toll, damage, ailment or other well-being impacts, property harm, loss of livelihoods, social and financial interruption, or ecological harm. Biological disasters may be in the form of:

- Epidemic influencing an excessively vast number of people inside a populace, network, or locale in the meantime, cases being cholera, plague, Japanese encephalitis (JE)/acute encephalitis syndrome (AES); or,
- Pandemic is a plague that spreads over an extensive locale, that is, a landmass, or even worldwide of existing, developing or reemerging infections and diseases, illustration being influenza H1N1 (Swine Flu).

Causes

Common Outbreaks

Natural outbreaks of ailment may progress toward becoming pandemics and accept unfortunate extent if not contained in the underlying stages.

Utilisation of Biological Agents by Terrorists

Bio-fear based oppression can be characterised as the utilisation of organic operators to cause demise, handicap or harm principally to individuals. Subsequently, bio-psychological oppression is a technique for fear-based oppressor action to win mass frenzy and moderate mass losses.

Bio-fear mongering is in this manner deliberate utilisation of natural specialists to cause ailment or passing through dispersal of smaller scale living being or poisons in sustenance or water or creepy crawly vector or by airborne to hurt human populace, nourishment yields and animals.

The three essential gatherings of organic operators, which could be utilised as weapons, are micro-organisms, infections, and poisons. Most natural specialists are hard to develop and keep up. Numerous separate immediately when presented to daylight and other ecological components, while others, for example, *Bacillus anthracis* spores, are seemingly perpetual.

Effect of Biological Disasters

Biological disaster forces overwhelming requests on the national human services framework and it will be the general well-being framework that will be called upon to deal with the outcomes. A compelling general well-being framework with part of a solid infection reconnaissance instrument, offices for quick epidemiological and research facility examination, proficient therapeutic administration and data, training and correspondence (IEC) are fundamental capacities for countering organic fiascos.

Potential Specialists in Organic Fighting

The potential specialists which might be utilised by fear-based oppressors could run from pathogens like *Bacillus anthracis*, *Yerseinia pestis* (Plague) and so onto living beings, for example, veriola (little pox) that have been confirmed as universally killed. Organic poisons or hereditarily altered pathogens could likewise be utilized. Fear-based oppressors may utilise new operators, or utilise living beings, for example, medicate safe or hereditarily built pathogens.

Mitigation

The basic insurance against regular and fake episodes of malady (bioterrorism) will incorporate the improvement of instruments for provoke identification of early flare-ups, seclusion of the tainted people and the general population they have been in contact with and preparation of investigational and restorative countermeasures. On account of purposely created flare-ups (bioterrorism) the range of conceivable pathogens is thin, while common flare-ups can have an extensive variety of creatures. The component required be that as it may, to confront both can be comparative if the specialist cooperatives are sufficiently sharpened.

- Reinforcing of incorporated reconnaissance frameworks in view of epidemiological studies; identification and examination of any illness episode.
- Foundation of early warning system (EWS).
- Coordination between general well-being, restorative care and insight organisations to forestall bioterrorism.
- Quick well-being appraisal and arrangement of lab bolster.
- Organisation of general well-being measures to manage auxiliary crises as a result of natural debacles.
- Inoculation of specialists on call and satisfactory amassing of vital immunisations.
- Recognising foundation requirements for detailing moderation designs.
- Giving fundamental learning of organic debacle administration through the instructive educational module at different levels.

FIRE

Fires are the mishaps which happen most regularly, whose causes are the most differing and which require intercession strategies and methods adjusted to the conditions and needs of every incident.

Depending on the type of fires (nature of the material on fire), meteorological conditions (wind) and the viability of the mediation, material harm can be constrained (a solitary auto, building or generation or capacity distribution centre establishment), or influence wide zones (woodland or agrarian flames, hydrocarbons, gas or other very combustible items, stockpiling or channeling establishments, harbor establishments and rail or marine transport hardware). Blasts are in an alternate classification.

Fires are a characteristic and useful component of many timberland scenes, yet they are hazardous when they happen in the wrong place, at the wrong recurrence or at the wrong seriousness. Every year, a huge number of sections of land of timberland around the globe are obliterated or debased by flame. Fire is frequently utilised as an approach to clear land for different uses, for example, planting crops. These flames not just change the structure and synthesis of backwoods, yet they can open up timberlands to invasive species, undermine organic assorted variety of the general population who live in and around the woodlands.

This is particularly the case for protect and fire annihilation on motorways, structures intended to be utilised by an awesome number of individuals (clinics, inns, silver screens, elevated structures, retail chains, etc.); fires influencing smokestacks, clothing types, cotton (bunches, free, dangerous residue), grub (maturation), fires in high stockrooms, storehouses or underground carports and additionally woods fires.

Every one of these sorts of mediation are liable to uncommon measures.

Preventive and Defensive Measures

Fires can spread pretty much quickly relying upon their causes, the nature of the material and merchandise land, the fire aversion establishments (programmed sprinklers), meteorological conditions, the manners in which the populace is educated and the activity it appears, and in addition the speed and productivity of the mediating administrations and of their putting out fires hardware.

Characterising, and controlling the execution of, the specific tenets of assurance against flames, particular to every venture exhibiting a potential peril, including the preparation of security staff, is additionally pertinent in this unique situation.

The prevention measures are
- Creating an observation, counteractive action and alert security at local levels.
- Arranging and planning (occasional upkeep) for putting out fires fighting

through satisfactory finishing of the region and suitable woods development restricting flame spread (rotating vegetation, freedom, trimming), making and keeping up get to ways (termination) and fire-break territories and additionally putting out fires gear, for example, water supplies (conductors, reservoirs), watch towers and meteorological posts, and the development of helicopter landing cushions.

- Observation and recognition of fires immediately when forecasted by the meteorological department.
- When the risk of fire builds, actuating an alert arrangement (essential mediation design) requiring the commitment of preventive intercession squads (fire fighters), and their wide situating as close as conceivable to the debilitated zones, and making accessible water planes and concentrated flying machines good to go.
- Readiness and concretisation (association) of a mediation system: This requires the setting up of specific administration programs guaranteeing the coordination of ground-breaking and proficient hardware and means for battling timberland fires (direction).
- Readiness administration and the coordination of the utilisation of the methods for mediation of the experts and the data and caution administrations for the populace require a safe transmission arrange (radio system).
- Arranging the clearing of the populace conceivably under risk in the different touchy territories, especially if there are dangers of blast (repositories and gas courses explosives or ammo dumps, hydrocarbon creation, dealing with or transport establishments, different perilous material, and so forth).

Guidelines for the Population

General precautionary measures and well-being measures identifying with a potential risk.

- Keep matches and lighters out of the span of youngsters and encourage them alert around flames and inflammable articles.
- Do not keep inflammable items (liquor, oil, gas compartments, paper, fabric, dried vegetable issue, and so forth) close to any wellspring of warmth.
- Know the directions identifying with flames, get some answers concerning security measures, know the whereabouts of gas and power conductors and figure out how to utilise residential putting out fires gear (dousers, fire reels and hoses, spouts, and so forth).
- Do not smoke, do not light flames, do not switch on electrical hardware or apparatus prone to make sparkles when dealing with, or pouring inflammable or lethal items (oil, liquor, gas, etc.), or in the event that they are spilling.
- Know the phone quantities of the putting out fires and common assurance administrations and of the police.
- Regard directions disallowing staying, lighting flames, or smoking in woodlands, manors, horticultural establishments, wooden houses, and so forth.

During Fire

- Avoid panic.
- Act in a quiet and astute way.
- Call for help by first cautioning the fire fighters (fire benefit) and unequivocally distinguishing the area (locality, street, number, kind of mishap, and furthermore the name and address of the guest).
- Immediately caution people in risk and those in charge of security in the building or the endeavour, particularly out in the open spots.
- Attempt to rescue people and animal in danger (wrap individuals whose attire is land in covers or coats and move them on the ground).
- Prevent the surge of air by shutting all entryways and windows and turning off ventilation.

- Do not utilise the lifts, leave the premises (stairs, ways out and crisis exits).
- If stair wells and passageways are loaded up with smoke, remain in the flat, close the entryway and water it every now and again, draft-confirmation it with wet clothes. Demonstrate your essence at the windows (without opening them).
- In the event that you are in a place that is getting loaded up with smoke, remain low on the ground where the air stays new.
- Battle the fire with every accessible mean (fire dousers, in-house hydrants, pouring water from utensils utilising the bath or sink as an ad libbed water store.
- Advise and manage fire fighters or different rescuers and take after their directions.

After the Fire

- Leave the house just if all parts of your body are ensured (cowhide shoes, gloves, cap, garments made of non-manufactured material).
- Assess your home and smother those parts which are burning (entryways, screens, and so forth).
- Assess the rooftop, the timber outline, the upper room and quench the soot which may have penetrated under the rooftop tiles and little openings by utilising the water hose or different beneficiaries loaded up with water.
- Water the vegetation encompassing your home and douse little flares assuming any.
- Help your neighbours and people in threat (emergency treatment).
- Obey requests of the fire fighters and of the experts' delegates.

OIL FIRE

Oil fires are more difficult to quench than consistent fires because of the tremendous fuel supply for the fire. In battling a fire at a wellhead, normally high explosives, for example, dynamites, are utilised to make a shockwave that pushes the consuming fuel and oxygen far from a well. The flame is evacuated and the fuel can keep on spilling out without bursting into fire.

In blowing the fire, the wellhead must be topped to stop the stream of oil. Amid this time, the fuel and oxygen required to make another inferno are available in abundant amount. At this risky stage, one little start (may be from a steel or iron instrument striking a stone) or other warmth source may re-touch off the oil.

To avoid re-start, metal or bronze devices, which do not strike flashes, or paraffin wax-covered devices are utilised amid the topping procedure. Fastidious consideration is utilised to keep away from warmth and sparkles, or some other start source. Re-start at the wellhead may appear as to a great degree ground-breaking blast, potentially much more terrible than the first victory.

With the latest technology and natural concerns, numerous straight forward oil fire today are capped while they consume.

There are a few procedures used to put out oil fire, which shift by assets accessible and the attributes of the fire itself.

Effects

Oil fires can cause the loss of a large number of barrels of raw petroleum everyday. It effects ecological imbalance caused by a lot of smoke and unburnt oil falling back to earth and also can cause tremendous financial loss.

Smoke from consumed unrefined petroleum contains numerous synthetic substances, including sulphur dioxide, carbon monoxide, residue, benzopyrene, poly sweet-smelling hydrocarbons, and dioxins (Hoobs et al, 1992). Exposure to oil fire can cause of the gulf war disorder.

Prevention Measures

- Dousing with abundant measures of water. Splashed from powerful hoses at the base of the fire.

- Using a gas turbine to shoot a fine fog of water at the base of the fire. Water is infused behind the fumes of the turbine in vast amounts.

- Using explosive such as dynamites to blow out the fire by compelling the consuming fuel and oxygen with smoldering heat from the fuel source. This was one of the most punctual viable techniques is still broadly utilised.

- Dry chemicals (for the most part purple K) can be utilised on small fires.

- Special vehicles called "Athey wagons" and also the common bulldozer ensured by layered steel sheeting are typically utilised in the process.

- Raising the tuft-putting a metal packaging 30 to 40 feet high over the wellhead (hence raising the fire over the ground). Fluid nitrogen or water is then constrained in at the base to lessen the oxygen supply and put out the fire.

- Drilling alleviation wells into the delivering zone to divert a portion of the oil and make the fire littler. (Notwithstanding, most help wells are utilised to pump overwhelming mud and bond profound into the wild well.) The primary alleviation wells were penetrated in Texas in the mid-1930s.

- Nuclear blasts for the national economy, the utilisation of underground atomic blasts were effectively utilised in the previous Soviet Association to stop well flames, the high warmth of the explosion at the same time dislodges and dissolves the stone in its region, and with that seals the beforehand penetrated hole (Nordyke, 2000; Broad, 2010).

COAL FIRE

The term "coal fire" refers to a consuming or seething coal crease, coal storage pile or coal waste pile. Two noteworthy causes have been recognised for coal fires.

Common Causes

Coal crease or residue can be uncovered by the disintegration or a subsidence occasion, strike by lightning or touch off by a rapidly spreading fire.

Human Causes

Grinding, power or oxygen can touch off coal peat or residue amid standard extraction, illegal mining, and transportation.

Coal fires consuming the world over are a natural calamity described by the emanation of harmful gases, particulate issue, and buildup side-effects. Underground mine flames and burning culm banks touched off by characteristic causes or human blunder are in charge of climatic contamination, corrosive rain, risky land subsidence, the decimation of botanical and faunal territories, human fatalities, and expanded coronary and respiratory maladies. A portion of the most seasoned and biggest coal fires on the planet happen in China, the United States, and India. Strategies used to battle coal fires incorporate slurry and powder infusion, surface and passage fixing, fluid froth innovation, remote detecting, and PC programming. Slippery, unusual, or cost restrictive coal flames may consume inconclusively, gagging the life out of a network and its environs while devouring a profitable characteristic asset.

Environmental Impact

Other than demolition of the influenced zones, coal fires regularly emit lethal gases, including carbon monoxide and sulfur dioxide. China's coal fires, which devour an expected 20–200 million tons of coal a year, make up as much as 1% of the worldwide carbon dioxide discharges from fossil fuels (Kevin, 2005).

A standout amongst the most unmistakable changes will be the impact of subsidence upon the scene. Another neighbourhood natural impact can incorporate the nearness of plants or creatures that are supported by the coal fire. The predominance of generally non-local plants can rely on the fire's length and the measure of the influenced region. For instance, close to a coal fire in Germany, numerous Mediterranean bugs and arachnids were recognised in a district with cool winters, and it is trusted that lifted ground temperatures over the flames allowed their survival.

Extinguishing Coal Fires

Keeping in mind the end goal to flourish, a fire requires fuel, oxygen, and warmth. As underground flames are exceptionally hard to reach specifically, firefighting includes finding a proper approach which tends to the association of fuel and oxygen for the particular fire being referred to. A fire can be confined from its fuel source, for instance through firebreaks or flame resistant obstructions. Numerous flames, especially those on soak slants, can be totally uncovered. On account of close surface coal crease fires, the deluge of oxygen noticeable all around can be hindered by covering the zone or introducing gas-tight hindrances. Another plausibility is to prevent the outpouring of burning gases so the fire is extinguished by its own fumes exhaust. Vitality can be evacuated by cooling, normally by infusing a lot of water. In any case, if any staying dry coal ingests water, the subsequent warmth of retention can prompt re-start of an once-extinguished fire as the zone dries. In like manner, more vitality must be evacuated than the fire produces. By and by these techniques are consolidated, and each case relies upon the assets accessible. This is particularly valid for water, for instance in bone-dry districts, and for covering material, for example, loess or earth, to anticipate contact with the climate.

FOREST FIRE

The most widely hazard in forest is forest fire. Forest fires are as old as the forest themselves. They represent a danger to the forest as well as to the whole administration to fauna and greenery genuinely irritating the bio-decent variety and the biology and condition of an area. Amid summer, when there is no rain for quite a long-time, the forest wind up covered with dry senescent leaves and twinges, which could blast into flares touched off by the scarcest start. The Himalayan forest, especially, Garhwal Himalayas have been burning frequently amid the last couple of summers, with colossal loss of vegetation front of that region.

Forest fire causes lopsided characteristics in nature and imperils biodiversity by diminishing faunal and botanical wealth. Traditional strategies for flame anticipation are not preventive and it is now raising public awareness on the issue, especially among those individuals who live near or in forest zones.

Causes

Common causes include lightning and drought but forest fire may also be started by human negligence or arson. Forest fires are also caused by normal causes and man-made causes.

- Natural causes: Many forest fires begin from regular causes, for example, lightning which set trees ablaze. Be that as it may, rain quenches such flames without causing much harm. High climatic temperatures and dryness (low moistness) offer good condition for a fire to begin.
- Man influenced causes: Fire is caused when a wellspring of flame like bare fire, cigarette or bidi, electric start or any wellspring of start comes into contact with inflammable material.
- Graziers and gatherers of different woodland items starting little flames to acquire great nibbling grass and additionally to encourage social occasion of minor forest products like blooms of *Madhuca indica* and leaves of *Diospyros melanoxylon*.
- The hundreds of years routine with regards to moving development (particularly in the North-Eastern locale of India and in parts of the conditions of Orissa and Andhra Pradesh).
- The utilisation of fires by villagers to avert wild animals.

- Surface fire is the most common type that burns along the floor of a forest, moving slowly and damaging trees.
- Fires began inadvertently by dispose of cigarette or bidi butts.

Classification of Forest Fire

There are three types of forest fires:
1. Surface fire: Common type of fires that burn along the base of a forest spreading slowly as the surface litter (senescent leaves and twigs and dry grasses and so forth) on the woods floor and is overwhelmed by the spreading blazes.
2. Ground fire that normally starts by lightning and burns on or below the floor of the forest.
3. Crown fire spreads up quickly by warmed air and incline tends to jumping along the top of the trees.

Impact of Forest Fire

Fires adversely affect the human and livestock in terms of socially, economically, ecologically, etc. including:
- Loss of important timber assets
- Degradation of catchment zones
- Loss of biodiversity and elimination of plants and animals
- Loss of natural lifeliving space and consumption of untamed life
- Loss of common recovery and decrease in forest cover
- Global warming
- Decrease of carbon sink asset and increase the level of CO_2 in air
- Drastic change in the climate of that region with undesirable living conditions
- Causing more soil erosion
- Ozone layer depletion
- Health problems
- Loss of livelihood of people who are dependent on collection of non-timber forest product to generate their livelihood.

Fire Management

Forest fire mitigation refers to prevention measures of reducing the risk of fires as well as dominating its severity and rapidly spread. Effective preventive measures allow administrators and supervisors to manage air quality, maintain ecological balance protect resources, and to limit the future incidence in forest.
- Vigorous follow-up activity.
- Introducing a forest fuel alteration framework at key focuses.
- Firefighting assets.

Everyone of the above segments assumes a vital part in the achievement of the whole arrangement of flame administration. Special emphasis is to be given to research, training, and development.

AIR POLLUTION

Air pollution alludes to the arrival of toxins into the air that are impeding to human well-being and the planet in general.

Causes of Air Pollution

1. **Natural sources:** The natural sources of air pollution are poisonous gases (SO_2, H_2S and CO, etc.) releasing from volcanic eruptions, forest fires, natural and organic decays or vegetative decay, marsh gases, blowing of sands and dust, extra terrestrial bodies, cosmic dust, pollen grains of flowers, soil debris, etc.

2. **Increase in population:** The rapid growing of population is one of the factors of air pollution. The increases in population worsens the environmental hazards and also contribute loss/hamper of forest cover and wildlife.

3. **Burning of fuels and fires:** About 90% of the energy used in homes and industries is from coal, gas and oil. Burning of fuel is responsible for air pollution.

4. **Emission from vehicles:** The automobiles such as cars, motors, taxies, bus, tracks, etc. release poisonous gases (CO, NO, hydrocarbon). Smog release from these automobiles is very toxic to environment.

5. **Use of pesticides in agriculture:** Different types of pesticides, insecticides, herbicides are used in field, causes air pollution because some amount of these substances is carried out by wind in different places, during application to the crops and field.

6. **Industrialisation:** A large number of industries are responsible about 20% of air pollution. The pollutants (CO_2, CO, SO_2, NO, NO_2, H_2S) are inorganic and organic gases and material in the smoke they produce. All these gases are injurious to health.

Impacts of Air Contamination

1. Air pollution cause deleterious effects on living organisms and may bring about death or sub-lethal pathology of kidney, lungs, liver, brain, etc.

2. Smog and soot come from cars, trucks, factories, power plants, engines—anything that combusts fossil fuels such as coal, gas, or natural gas. The tiniest airborne particles are dangerous because they can penetrate the lungs and bloodstream and worsen bronchitis, lead to heart attacks, and even hasten death.

3. Air pollutants such as mercury, lead, dioxins, and benzene most often emitted during gas or coal combustion, incinerating, or in the case of benzene cause eye, skin, and lung irritation blood disorders, nervous, and endocrine systems, as well as reproductive functions.

4. Emission of greenhouse gases lead to warmer temperatures and climatic change such as rising sea levels, more extreme weather, heat-related deaths, and increasing transmission of infectious diseases like Lyme.

5. Pollen and mold from trees, weeds, and grass are also carried in the air, are exacerbated by climate change, and can be hazardous to health.

Prevention and Control of Air Pollution

1. Use of wood and dung cakes should be used instead of biogas, kerosene or electricity.

2. Use of liquefied natural gas (LNG) in industries is beneficial to environment.

3. Using environment-friendly industrial processes so that emission of pollutants is minimised.

4. Installing devices such as filters, electrostatic precipitators (ESPs), inertial collectors, scrubbers reduce release of pollutants.

5. Increasing the height of chimney.

6. Closing or shifting industries which pollute the environment, away from populated areas.

7. Development and maintenance of green belt of adequate width.

8. The emission standards for automobiles have been set which will reduce the pollution.

WATER POLLUTION

Water, the most abundant and wonderful natural resources, is extremely essential for survival of all living organisms. Water pollution is one of the best emergencies confronting the nation. Nowadays, fresh water has become a precious things and its quality is threatened by numerous sources of pollution which are as follows:

1. **Sewage and domestic wastes:** About 70% of water pollution is caused by sewage, domestic wastes and food processing plants. It also includes human excreta. Soap, detergent, rubbish, garden waste, sewage sludge, etc.

2. **Industrial effluents:** Industrial effluents discharged into water bodies contain toxic chemicals, hazardous compounds and pollutants from numerous industries. These pollutants when discharged from sewage system, pose several pollution problems.

3. Pesticides, fertilizer, insecticides, herbicides, agricultural wastes, plant and animals debris are reported to cause heavy pollution to water sources. These are washed off through rainfall, irrigation and drainage into water bodies, where they severely disturb the aquatic ecosystem.

4. A large number of huge amounts of garbage streaming out of our urban areas and towns discover their way in waterways. As chemical fertilisers and dungs are being utilised for cultivating as well, water sources are getting seriously undermined.
5. Decline in the water quality flowing coursing through the fields.
6. Social and religious ceremonies, for example, skimming dead bodies in the water, washing, littering.
7. The oil slicks from ships.
8. Acid rain.
9. Global warming.
10. Eutrophication (the consumption of oxygen in a water body, which murders amphibian creatures).

Mitigation Strategies

1. Water is polluted due to soil erosion. So, if we conserve soil then can stop water pollution to some extent. The way to stop soil erosion is planting more plants or trees. We can adopt such methods of cultivation that improve the health of the soil rather than the spoil it.
2. Remove nutrients, disinfect for removing pathogenic bacteria, and aeration removes hydrogen sulphide and reduce the amount of carbon dioxide and make water healthy and fit for aquatic organisms.
3. Adopting the correct methods of disposal of toxic waste is also extremely important. In the beginning, we should reduce or not use such products that include harmful organic compounds.
4. Oil spill out of cars or machines is also one of the factors of water pollution. Cars or machines should be regularly checked to ensure that there are no oil leak.
5. Water requirement should be minimised by altering the techniques involved.
6. Water should be reused.
7. The quality of waste water discharge should be minimised.
8. Cleaning of waterways and beaches.
9. Stop using biologically non-degradable materials such as plastic.

Water pollution has now taken the form of an emergency. So, we need to take big steps urgently. If we want that our citizens continue to get safe drinking water and water sources remain safe for a long duration, we will have to take steps for it from today itself. The delay can prove to be fatal.

DEFORESTATION

Forests cover 31% of the land region on our planet. They give us oxygen and give homes to individuals and natural life. Huge numbers of the world's most undermined and jeopardised creatures live in woods, and about 1.6 billion individual depends livelihood non-timber forest product. However, forests around the globe are under risk from deforestation, endangering these benefits. Deforestation comes in various forms such as fires, cutting for trees, farming and unsustainable logging for timber, and degradation because of environmental change. This effects individuals' occupations and debilitates an extensively plant and animal species. We are losing 18.7 million acres of forest yearly, proportionate to 27 soccer handle each moment.

Forests play a crucial role in prevention/mitigation of climate change since they go about as a carbon sink—soaking up carbon dioxide that would some way or another be free in the air and add to continuous changes in atmosphere. Deforestation undermines this critical carbon sink function. It is reported that 15% of greenhouse gases emission are the after-effect of deforestation.

Deforestation can happen rapidly, for example, when a fire clears through the scene or the timberland is obvious to clear a path for an oil palm estate. It can likewise happen step by step because of continuous woods debasement as temperatures ascend because of environmental change caused by human action. While deforestation has all the earmarks of being on the decrease in a few nations, it remains stunning high in others—including Brazil and Indonesia—and a grave danger to our reality's most important timberlands still remains.

Causes

There are numerous reasons for deforestation. About half of the trees illicitly expelled from forests are utilised as fuel. Some other regular reasons are:

1. To make more land accessible for lodging and urbanisation.
2. To reap timber for business purposes, for example, paper, furniture and homes.
3. To make fixings that are exceptionally prized purchaser things, for example, the oil from palm trees.
4. To build room for ranching.
5. Main causes for deforestation are burning trees and clear cutting. These strategies leave the land totally fruitless and are disputable practices.

Impact of Deforestation

1. **Flood and droughts:** Deforestation prompts disintegration of land erosion that the trees assume a vital job in keeping up the surface of the mountains and cause characteristic obstructions to the quickly rising precipitation water. Subsequently the water level of the waterways increments all of a sudden, causing surges.
2. **Air pollution:** There are grave outcomes for forests obliteration. Its greatest burden is as air contamination. The air where there is absence of trees gets polluted. Therefore people suffer from ill effects of numerous maladies, particularly breathing issues, for example, asthma.
3. **Loss of soil fertility:** The cow dung and vegetative residues are used for crop production. After some time the float of this nourishment impacts the efficiency of the dirt, it causes debasement of soil fertility.
4. **Elimination of species:** Because of the destruction of forest, untamed life is vanishing. Numerous species have vanished.
5. **Global warming:** The major cause of global warming is deforestation, which directly affects the common environmental change,

by this increasing the worldwide tempe-rature.

6. **Expansion of deserts:** Because of consistent reduction in the forest, and thereby erosion of the land, the desert is spreading on a major scale.
7. **Impacts of industrialisation:** Trees and plants keep the earth from being polluted by keeping those harmful gases from dissolving in the climate, and keeping the particles of fiery remains and sand and so on from rising as well. These days, there is a surge of ventures in the urban com-munities, even towns and towns.
8. **Harm to ozone layer:** The ordinary condi-tion of the earth because of deforestation has turned out to be dirtied. It is presenting grave threat to the ozone layer, which is fundamental for the general resistance of the earth.
9. **Loss of livelihood:** Many people generate income from forest product. Due of decre-asing the forests they loss their livelihood.

Prevention Measures of Deforestation

1. Planting more trees.
2. Quit printing and go paperless. Use personal computer or electronic devices to records files or papers and avoid printing.
3. Moving towards buying recycled products. By augmentation, you are proceeding with your proactive exercise in redirecting the interest for clearing land.
4. Restoring the ecosystem provided by forests including carbon storage, water cycling and wildlife habitat.
5. Reducing the release of carbon dioxide in the atmosphere.
6. Rebuilding wildlife habitats.
7. Build up parks to protect rain forest and wildlife.

INDUSTRIAL WASTEWATER

Industrial wastewater is one of the vital contamination sources in the pollution of the water condition. Amid the most recent century a colossal measure of modern wastewater was released into rivers, lakes and seaside regions. This brought about genuine contamination issues in the water condition and made negative impacts the eco-system and human's life. There are numerous sorts of industrial wastewater in view of various industries and pollutants; every division creates its own specific blend of poisons. Like the different qualities of industrial wastewater, the treatment of modern wastewater must be composed particularly for the specific sort of pollutant delivered.

Types

There are various types of industrial wastewater on the basis of different industries and pollutants. Table 2.1 shows that each sector/ industry produces its own particular combination of pollutants.

Inorganic Industrial Wastewater

It is produced in the iron and steel industry, non-metallic mineral industry and commercial industry for surface possessing of metals. This wastewater contains huge amount of suspended materials which are eliminated by sedimentation, chemical flocculation through the addition of iron, salts, agents and polymers.

Organic Industrial Wastewater

Organic wastewater is produced by paper, leather factory, factories in which cosmetics, soaps, detergents, pesticides, herbicides are produces, textiles, oils, fermentation and metals factory. This contains organic industrial waste flow from chemical industry, which mainly use organic substances for chemical reaction. These can be removed only be pretreatment of wastewater followed by biological treatment.

Impacts of Industrial Wastewater

The wastewater produced in different mechanical procedures can essentially realize the accompanying changes when they are filled the water bodies.

1. **Effect on environment:** Mechanical water contamination can have extensive consequences for the biological system. The water utilised in different mechanical procedures interacts with harmful synthetic substances, overwhelming metals, natural muck, and even radioactive ooze. Along these lines, when such contaminated water is tossed into the sea or other water bodies with no treatment, they end up unfit for any human and horticultural utilise.

2. **Thermal pollution:** The radioactive slime stored at the base of water bodies can remain exceedingly radioactive for a long time, and posture genuine well-being dangers for individuals living close-by. Atomic reactors are additionally a noteworthy wellspring of warm contamination alongside control plants. Warm contamination

Table 2.1: Water pollutants by industrial sectors	
Industry	*Pollutants*
Iron and steel	Oil, metals, acids, BOD, COD, phenols and cyanide
Textile and leather	BOD, solids, sulphates, chromium
Pulp and paper	BOD, COD, solids, chlorinated organic compound
Petrochemicals and refineries	BOD, COD, mineral oils, phenols, chromium
Chemicals	COD, organic chemicals, heavy metals, SS, cyanide
Non-ferrous metals	Fluorine and SS
Microelectronics	COD and organic chemicals
Mining	SS, metals, salts and acids

BOD: Biological Oxygen Demand, COD: Chemical Oxygen Demand
Source: Shi, Hanchang, 2002.

alludes to an expansion in encompassing water temperature. It can adversy affect amphibian or marine life, as a few living beings are to a great degree touchy to slight changes in temperature.

3. **Impact of eutrophication:** At the point when the supplement substance of water experiences transforms, it can aggravate the sensitive equalisation of the biological community. For instance, when the supplement substance of water builds, which is known as eutrophication, it can advance algal blossom. Algal sprout can drain the oxygen substance of water. Despite the fact that green growth create oxygen in the daytime with the assistance of photosynthesis, during the evening, they utilise the oxygen broke down in water.

4. **Increment the murkiness of water:** Mechanical wastewater can build the dinkiness of water. This thus, can keep daylight from achieving the base of the water bodies. Thus, base abiding plants can neglect to photosynthesize. Intemperate dinkiness of water can likewise hinder the gills of fish, and in this way, make it troublesome for them to take up broke up oxygen from the encompassing water.

5. **Impact of synthetic substances:** The regular modern toxins in charge of causing water contamination are sulphur, asbestos, toxic solvents, polychlorinated biphenyl, lead, mercury, nitrates, phosphates, acids, alkalies, colors, pesticides, benzene, chlorobenzene, carbon tetrachloride, toluene, and unstable natural synthetics. Synthetic substances like sulphur is unsafe for marine life, while asbestos is known to be a potential cancer-causing agent. Drinking water debased with asbestos may build the hazard for favourable intestinal polyps.

The compound chlorobenzene found in bug sprays and colours can antagonistically influence the liver, kidneys, and the focal sensory system. Toluene is fundamentally a poison created by the oil and oil industry, which also can influence the liver, kidneys, and the focal sensory system. Unpredictable natural synthetic concoctions are essentially solvents utilised in an extensive variety of modern and family unit items. At the point when not arranged appropriately, these synthetics can dirty the groundwater. They can cause an extensive variety of medical issues like cerebral pains, sickness, liver harm, and memory impedance.

Preventive Measures

In spite of the fact that modern water contamination is hard to contain, it is not inconceivable. A more prominent mindfulness should be made among the normal mass about how water gets contaminated, its consequences for human well-being and marine life, and how it very well may be anticipated. It is not conceivable to diminish water contamination without open collaboration, and participation of modern units.

Strict contamination control laws and enactment, and their successful usage do have an imperative job in controlling any sort of contamination. The advancement of reasonable contamination control hardware and motivators from government for introducing such gear can urge businesses to consider up contamination control important.

Some mechanical squanders are customary squanders, much the same as household sewage. Such squanders can be effortlessly treated by the metropolitan offices. However, a few squanders like substantial metals, unpredictable natural mixes, and oil and oil do require extraordinary medicines. Enterprises can introduce a pre-treatment framework to separate such risky squanders. The incompletely treated wastewater can be sent to the metropolitan framework for encourage purging.

Substantial scale ventures create a lot of wastewater. Such ventures ought to upgrade their assembling procedures to decrease the measure of contaminations and work their own particular on location treatment frameworks. The treatment of modern wastewater should be possible in three phases: (i) Primary treatment that includes mechanical procedures, (ii) optional treatment by organic procedures, and (iii) tertiary treatment that should be possible with the assistance of natural, physical, and substance forms.

In essential treatment, contaminations are isolated from water with the assistance of screening, crushing, flocculation, and sedimentation forms. In optional treatment, natural techniques are utilised for wastewater treatment. At last, the wastewater is reused in tertiary treatment with the assistance of organic, physical, and substance forms. Warm contamination then again, can be controlled by shaping cooling lakes, or by utilising cooling towers.

ROAD, RAIL, AIR AND SEA ACCIDENTS

Significant air, rail, street, and oceanic accidents are by and large same as they include quick moving vehicles conveying numerous individuals or expansive amounts of products and substances that can make direct or indirect harm to people or environment in the site of the accidents. The impacts of these accidents are multiples when a few of the same, or extraordinary, sorts of vehicles or methods for transport are included or when vehicles hit structures or establishments protecting individuals or containing substances that are risky to man and nature (towns, stations, air terminals, distribution centers and so on).

The control of all accidents is the foremost the responsibilities of the chief and workforce of the influenced methods for transport. It is dependent upon them to restrict the subsequent harm as much as could reasonably be expected. Passengers must comply with the mandates of the staff on load up (defensive and protect measures) and behave as they are directed by the regulation on calamity circumstances, particularly air, rail or sea debacles.

Preventive and Defensive Measures

In perspective of the diverse means of transport, and subsequently the conditions in which the accidents took place, usually difficult to ensure the protection and rescue of the victims and people who are in danger. Subsequently, priority must be given to anticipation and thereafter to speed up the activities.

To this end the association of movement observation and traffic signal and warning systems and plans for mitigating the risk set up ahead of time by the security administrations of transport organizations, as a team with the neighbourhood, regional and national experts and interceding bodies.

General Precautions and Preventative Measures for Clients of Different Kinds of Transport

1. Informed yourself about the risks engaged with utilising different kinds of transport.
2. Concentrate the posters, presentations, public address system provided by the transport authorities.
3. If there should be an occurrence of immediate accidents, ensure that you know the whereabouts of the security and safeguard contraption.

During Accident

- Try to avoid panic and keep calm, encourage and help people and neighbours who are in trouble.
- Follow or obey commander's (or driver's) orders and those of the group.
- Keep identity card and imperative individual documents and others with you.
- When the accident or wreck happens, endeavour to free yourself from the lodge and to escape rapidly from the destruction, particularly if there is a threat of fire or risky breaks. Look for asylum at an adequate separation.
- Ensure the operational process by alerting the local authorities.
- In the event of a noteworthy mischance happening in a street with movement, coordinate the activity while dealing with your own well-being.

- Do not touch or move the truly injured except if there is a danger of fire or poisonous exhaust.
- It is the obligation of observers to alarm the save benefits and to informed them the exact area and nature of the accident, the sort of vehicle included, the qualities (code number) of any unsafe substances and the conceivable number of casualties. Witnesses ought to likewise give their names and addresses.

After the Accident

- Avoid panic and maintain a strategic distance from spot.
- Adhere to the directions of the authorities and of the save workforce.
- If possible, work together with the rescuers and with the legal specialists and specialists responsible for the investigation.

ROAD ACCIDENTS

Road accidents are the most incessant and the reason for the most harm. The purposes behind this are freedom given to the driver, reckless driving and densely road traffic. Accidents related to heavy loaded vehicles occur regularly despite call while, regard of the stacking directions and the highway Code, and additionally the commitment for drivers to adjust their speed, which influences stopping distance, to the activity and climatic conditions (rain, ice, mist, etc.). The aversion of road accidents is additionally critical and will be guaranteed by strict laws, by specialised and police controls, continuous preparing for drivers (particularly those associated with the vehicle of risky substances) and, if require be, by lawful and managerial punishments for those responsible.

Protection Measures

As road accidents, it is vital to make the difference between those happening on motorways or highway and those occurring on different streets, remembering the extraordinary directions that may oversee certain routes courses or sensitive areas.

Operational controls of accident on streets are at first operationalised by the traffic or rescue management team of the focal point of the motorway, which is cautioned by witnesses or by a police. If possible special disaster management team, specialist, health specialists can also be called upon to help. A special plan for the management will be executed. Road accidents other than the highway or motorway accidents go under the control of the political experts of the region in which it happen, who ensure the initial interventions utilising the methods available to them. If there should arise a terrible accidents a higher administration level will be called upon to intervene and direct the activities.

TRAIN ACCIDENT

Rail accidents happen when trains going on similar tracks collide or when trains derail because of technical fault specialised blames in the moving stock, the rails or the security personnel, or as a result of climatic hazards such as landslides, avalanches or terrorists attack. Such types of accident cause direct or indirect harm to individuals and nature, particularly when people carrying inflammable articles in the trains. This can only be avoided by prevention and protective measures such as establishing laws and regulations, trained personnel. Signal system, etc. that limit specialised and human disappointments and malevolent assaults. For example, Indian Railway passenger train crashed into a crowd of people in Amritsar, Punjab on 19 October, 2018. The accident occurred in the evening, killing about 59 people and 100 were injured (Fig. 2.1).

Protective Measures

If there should be an occurrence of rail mishaps, the alert, observation and rescue will be controlled by the regional traffic control centre of the rail or by the nearest station. The local authorities can take control of the

Fig. 2.1: Amritsar train accident, 2018

preliminary prevention by coordinating with the other authorities such as police, fire protection, doctors, etc. Besides these red alerts for rail accidents can be initiated by higher political experts. Security, research, safeguard and help measures will rely on the intervention strategies managed by the mischance and the methods accessible.

AIR ACCIDENT

Air accident can be of normal, specialised or human cause, for example, mechanical breakdowns, carelessness or fear-based oppressor assaults. Normally, small aircrafts such as helicopters, light planes, lightweight flyers do not cause any calamities, since the quantity of casualties and the effect of the accident is constrained. This is not the situation for substantial flying machine, for example, transport planes or fly warriors, despite the fact that mischances including this sort of air ship are generally uncommon; when they do happen, usually in the border of runways and in the pivot of take-off and runways. The most serious risk emerges from the radioactivity exuding from an atomic reactor fueling such a vector.

Intervention and Rescue Measures

Immediately after the air accident, in the area of airport terminal, airport authorities should control of the situation. They can, if require be, call upon neighbourhood fire detachments and common resistance fortifications. For the most part the special disaster rescue team may be initiated.

If the occurrence happens outside the zone under control of the airplane terminal specialists, the political experts of the influenced region will be responsible for the organisation of local intervention groups (police, fire protection, common assurance, health service, etc.). Control of the protect endeavours will, if important, be assumed control by higher specialists (local, departmental) as a team with specialists from the flight administrations and the legal (building up obligation regarding the debacle, distinguishing casualties, accumulation and compensation of individual things) and delegates of the vehicle organisation included.

SEA ACCIDENT

Oceanic catastrophes have enormously lessened, basically in view of enhanced boats and more navigation systems, yet most likely additionally as a result of the expanding and semi selective utilization of business flying for traveler transport, at any rate for worldwide travel. Along these lines, most mishaps currently happen when crossing ocean channels or estuaries and include over-burden or old vessels. Also, ferry boats have also been associated with imperative fiascos on account of their insecurity in terrible climate, specialised issues or navigational mistakes. The major environmental debacles have happened when petroleum tankers have steered into the rocks, in light of the loss of compartments of risky items or even in view of the consider dumping of these items (squander). It is essential to keep this to the degree conceivable by guaranteeing the regard of national and global directions through incessant and serious check up.

Intercession and Protect Measures

Safeguards on account of oceanic or stream calamities pursue the standards of solidarity between nautical individuals and the standards of national and worldwide sea law.

In case of an occurrence, the protect measures on board are the duty of the captain of the vessel whose need ought to be to save the travelers and afterward the group. In addition, oceanic countries for the most part have a national administration for the ocean and traversable channels which is incharge of guaranteeing the supervision and security of national waters and drifts and furthermore the association of safeguard adrift.

On account of the particular states of route adrift, and on lakes and waterways, interventions ought to be the question, in each state worried, special protection/measures plans should be put into action when the alarm is given by chief of operation.

In the event of ecological disasters (chemicals and radioactive substances, hydrocarbons), the protection measures for the coasts ought to be executed in close coordination with the United Nations Department of Humanitarian Affairs (UNDHA).

Questions

1. What causes man-made disasters and how do they affect us?
2. How do you protect yourself and your family in case of a nuclear disaster?
3. Differentiate Natural Disasters and Man-made Disasters with examples.
4. Describe in brief the causes of climate change.
5. What are the typical effects of deforestation?
6. What do you mean by volcanic eruption?
7. What is smog? What are the various reasons for the presence of smog in the Northern region of India during the month of November?
8. What are the hazardous effects of volcanoes?
9. What are the causes of global warming?
10. Describe the impact of global warming on climate related disasters.
11. Describe wild fire.
12. Describe oil fire.
13. What is sea level rise? Explain its cause and effects.
14. What are the main consequences of chemical disasters.
15. List down the mitigation measures of biological disasters.
16. Classify fire. What to do during and after the fires?
17. What are the major sources of outdoor pollution?
18. What helps protect humans from air pollutants?
19. How can we cut down or prevent air pollution?
20. Explain the traditional way of purifying water to make fit for drinking.
21. How we can reduce, reuse and recycle water?
22. "CO_2 contribute global warming" explain it.
23. List any two effects of industrial wastes on land.
24. What are the significant of BOD–COD ratio?
25. What is the toxic substance in industrial waste?
26. Why do certain locations have discharge requirements?
27. List the three alternatives of the disposal of industrial waste.
28. What is the significance of environmental legislation?
29. Discuss different types of transport accidents and prevention measures.

3 Disaster Management in India

DISASTER MANAGEMENT—EFFORTS TO MITIGATE NATURAL DISASTERS AT NATIONAL AND GLOBAL LEVELS

Damage and losses due to natural hazards can be minimised if effective, timely and adequate measures are taken in pre-disaster planning/preparedness, management during disaster relief by providing safe drinking water, food, shelter, emergency medical aid and also taking the measures required to contain epidemics and rehabilitating the people. This requires a holistic approach and highly coordinated efforts by different Government departments, and public, private and voluntary organisations and cooperation of the people in the areas. Disaster management/mitigation require multi-disciplinary efforts involving financial, scientific and technological inputs, managerial effectiveness, social understanding and above all dedicated efforts of all concerned agencies.

Mondal et al (2016) concluded from their study that disaster can be reduced considerably if appropriate pre-disaster measures are taken including effective and early communication of cyclone warnings and evacuation of people to safer places depending on the situation. There is a thumb rule that every dollar invested in disaster preparedness will save twenty times later on.

There needs to be a paradigm shift from erstwhile relief-centric and post-event response to a pro-active approach encompassing prevention, mitigation and preparedness-driven disaster management-declared by the National Disaster Management Authority in its flood management policy of 2008. The Government of West Bengal indeed maintains a Disaster Management Department but it reacts late as usual after the event when the farthest terrains are difficult to approach and the afflicted community is in deep danger. In view of predictions made in the fourth assessment report of the inter-governmental panel on climate change that there will be more extreme hydro-meteorological events affecting lives and livelihoods of the people living in Indian subcontinent, we need a pro-active disaster management plan.

One of the major causes for loss of human life due to natural disasters is that people live in disaster-prone areas and that of houses, buildings and other structures are not disaster proof. While it is practically not possible and economically feasible for any society, to afford or build houses and structures that are absolutely safe against worst disasters, the main aim of the scientific and technological efforts is to develop rational methodologies for minimising the damage.

To mitigate the effects of natural calamities, short-term strategies like relief and long-term programmes such as avenue plantations, construction of major and medium projects, soil and water conservation measures may serve to minimize flood and cyclone. Non-government organisations and the agencies associated with the activities like relief, employment, rehabilitation, technology generation and dissemination are to be included as an active partner to improve the situation in flood and cyclone prone areas.

Some salient features are emerging from the present study in respect of disaster management policy which are enumerated below.

Structural Measures

Strengthening Coastal Embankments and Infrastructure

Embankments along the coastline and around the offshore islands can prevent the movement of strong surges inland. The study area has over 3500 km of embankments built along its rivers and coastal areas. Coastal embankments serve to protect coastal communities and other productive resources (e.g. agricultural land) from tidal surges, but as a coastal defence mechanism, most have proved to be inadequate. This study suggested that repair and maintenance of the damaged embankments and communication system on a priority basis to protect the affected area from further flooding and tidal surges. Government and local institutions should take adequate measures to built concrete embankment and existing embankments should be maintained and monitored regularly.

Providing Basic Support Services

Among basic support services, drinking water is considered the most critical in study areas communities, where sources of safe drinking water may be contaminated through direct saline water intrusion as a result of tidal surges or indirectly through soil salinity. In most vulnerable communities, standards of sanitation and health services are still very low, with many households reliant on harvested rainwater as a major source of drinking water. This is problematic because the rainy season usually only lasts for a couple of months after which access to fresh drinking water deteriorates. To address this issue, greater efforts should be made to improve safe water infrastructure in at risk areas, including saline water treatment services. Additionally, the provision of (minimum standard) sanitation systems and health services would greatly improve the lives of those in affected areas. Local institutions and NGOs should take steps by buying and supplying essential commodities for the local markets to distribute the relief materials without any bias. The organisation was also collecting fund in order to be able to give the required and necessary help. This situation never happened before, and it was really difficult to deal with as the demand was huge.

Relief measure

i. Stocking of sufficient food materials for people and domestic animals.
ii. District/block medical/relief teams asked to take position at strategic points and coordinate with village volunteers/task forces.
iii. Set up temporary shelters/relief camps after initial warning.
iv. Supervise and monitoring relief measures.

Build up Disaster Resistance Houses

Housing is on the highest demand among the affected people. Most of the people lived in mud houses that are prone to risk. Because of cyclone and for water logging situation, infrastructural items have been collapsed almost all the shelters including safe sanitation system and this raised the vulnerability for people's livelihood. It was mainly occurred because of insufficient institutional and infrastructural supports for them. It changes their income opportunities. Government and NGOs should take initiative and motivate the people to build up disaster resistant houses. If only one room of every house, capable of accommodating the entire family during the few hours of emergency, may be designed to resist the wind and surge.

Shelter for Livestock

The cyclone has caused death of livestock. It is mainly due to crisis of potable water, feeds and fodder and isotonic dehydration. Construction of earthen mounds with the top level raised above the maximum height of surges is a possible solution for sheltering livestock. The major reason is that the mounds are located away from human settlement and people are reluctant to walk a long distance to take their cows, buffaloes, sheep or goat there. Moreover, lack of normal time use and maintenance leads to erosion of slopes and unwieldy growth of grass and shrubs, which make them unusable during emergency.

Supply of Agricultural Inputs

Agricultural production system was totally hampered after disaster due to high salinity and PH condition of soil. Paddy, wheat, sugarcane, chilli and pulses production was highly destroyed as observed from the present study. The ministry as well as the concerned officials of the forest department and agricultural department must be undertaken immediately for developing a detailed stepwise scheme with the aim of generating livelihood for the affected people. The concerned department must handover to saline resistance seedlings to the farmers, who were helpless as they could not germinate seeds as acres of land were submerged under saline water. If possible the government must supply of inputs at subsidised rate. The farmers will be trained to prepare seed bed and to adopt agricultural practices that are economically viable, environmentally sound.

Afforestation

Coastal afforestation has been intrusion of storm surges inland. In some cases, they may help in reducing the depth of inundation. The mangrove population must be saved first. Mangrove is the variety of trees that can grow in a mixed ecological situation of sweet and salt water typical of the Sunderbans. Along the river 30–40 ft of mangrove wall can save the village. The salinity tolerant tree species that can be grown in the disaster affected areas. Berry, pongamia, tamarisk tree, ber, guava, papaya, fig may be planted.

Building up Community Shelters

Construction of community shelters capable of providing victims to the population likely to be affected appears to be a feasible option. Community shelters are usually reinforced concrete framed structures, with brick masonry infill walls. The ground floor is left open to allow the surge water to flow through the building. The upper floor of these two or three-stored buildings are used as shelters. There was very few community shelters are in the study areas. Government should take initiative to build up community shelters in the vulnerable areas that protect the villagers to loss of lives.

Normal Time use of Shelters

The frequency of cyclone and surges is such that the shelters are likely to be used only once in 4–5 years. Since the shelters buildings are quite expensive structures, it would not be economically justifiable unless some normal time use can be ensured. Moreover, scientists showed that unless these buildings are used regularly, it is very difficult to maintain the facilities.

The normal time use of shelters may include the following:

- Educational institutions (e.g. primary school, secondary school, college)
- Vocational training institutions
- Health and Family Welfare Centre
- Community Centre
- Offices
- Passenger Terminals

Non-structural Measures

Strengthening the Early Warning System

It is seen that there was no weather forecasted system in the study areas. It has been proven by researchers/scientists that early warning systems and access to cyclone shelters have helped to reduce the number of deaths from cyclones. Building on these successes, early warning systems should be scrutinised to identify options for further improvement, with the potential impacts of climate change on extreme events in mind. For example, awareness building activities should be undertaken to ensure that at-risk communities understand how the system works, and what to do when an early warning is issued. Additionally, this mechanism should also seek to provide early warning system well in advance, before 3–4 days of occurrence of cyclonic storms to the vulnerable groups who can prepared themselves from risk and lesser will be the damage.

Strengthening of weather forecasting and warning systems through application of sophisticated technologies like satellite remote sensing, computer-based mathematical watershed modeling and telemetric methods of real time data collection and transmission, etc. are essential for meeting the challenge posed by flood hazard. Geological Information System (GIS) provides an effective tool for management of floods.

- Construction of embankments and dikes
- Widening or deepening of channels to increase the carrying capacity of rivers
- Construction of diversion of channels to carry excess flood water
- Construction of dams to store flood waters.

The measures reflected in this section are essential for the betterment of the people staying in disaster affected areas. It is therefore suggested that the central government and district administration take appropriate steps for mitigation of the disasters in affected areas and more particularly in the vulnerable areas of flood and cyclone.

Forecast and Warning System

Most of the people in the study areas belonged to low social strata. The available information should be effectively transmitted to the communities. Mass media plays pivotal role for transmitting the messages. The role of communication technology in disaster management is to keep the flow of real-time data and information during all these phases. Dissemination of warning messages which are clearly understood by the local people appears to be a more difficult task. The traditional system of warning signals originally developed for maritime use led to considerable confusion and is now being replaced by a more easily understood numbering system.

Supporting Long-term Income Generation Activities

Following the aftermath of cyclone, much of the recovery effort focused on short-term and periodic relief activities (e.g. providing food, shelter, and drinking water, etc.). In anticipation of future cyclones and other extreme events, additional long-term recovery measures are needed to help restore and maintain livelihoods and adapt to the lasting impacts of these disasters.

In this context, long-term support is needed for capacity building in cyclone-affected communities to help diversify livelihoods and facilitate self-recovery measures. Feasible options to support income generating activities include support for agricultural production, small-scale aquaculture development, creation of small and medium-size businesses and other forms of employment through access to credit and vocational training. Local farmers can be supported with agricultural inputs, such as saline tolerant seeds, and interest-free credit to increase the local production base and strengthen local food security.

Training and Deployment of Local Institutions and Volunteers

Members of local institutions and volunteers have to be trained to disseminate the warning among the people, to assist the people to move to shelters, to carry out search and rescue operations immediately following a disaster and to provide first aid to injured people. These members undergo regular training and have contributed significantly in reducing loss of human lives.

Public Awareness

A programmed have to be initiated to raise the level of awareness of the vulnerable communities. Actions to be taken by individuals and families before, during and following disasters are discussed with the community and training imparted to community leaders. Theatre groups go round village staging plays based on preparedness measures and what to do during cyclones. Measures have been initiated to include topics related to disasters in primary and secondary school curricula.

Planning

The primary consideration in locating a shelter is to ensure that it is easily accessible to potential victims. Study showed that people tend to move to the shelters only a few hours before the water starts rising, by which time the wind speed may be quite high. Study also showed that people who owned livestock do not like to take them to a shelter, which is not available to the study areas. Construction of network of shelters would prevent the loss of human lives and livestock during future disasters.

Short-term Rehabilitation

Major short-term rehabilitation activities included food security, creation of community assets, reviving schools, social mobilisation and group formation, etc. Government and local NGOs should be initiated Food for Work Programmes in affected villages to provide food security and facilitate restoration/construction of community assets such as water sources, irrigation facilities such as canals and earthen check-dams, roads, and other civic infrastructure.

Long-term Rehabilitation

Local institutions that could mobilise resources went ahead with long-term rehabilitation initiatives even as others withdrew from the scene. Restoration of farm and non-farm livelihoods, construction of schools-cum-cyclone shelters, and initiatives to strengthen community-based disaster preparedness would be highlights of rehabilitation efforts made by local institutions.

Efforts should be Comprehensive and Inclusive, and should Promote Multi-organizational Participation

Disasters require inter-organisational coordination and cooperation for an effective response; therefore preparedness efforts should include all of the groups responsible for the various emergency management functions. Preparedness efforts should include representation from emergency management, law enforcement, fire, city management, public health, citizen and voluntary groups, schools, nursing homes, hospitals and health care organisations, the business community, and other sectors in order to create a network of organisations to support essential functions in a disaster event.

It is important to devise preparedness strategies that are intentionally broad in part because of the tendency for preparedness activities to be vertically integrated—or stove piped—rather than horizontally integrated, across community organisations and sectors. Sector-based preparedness efforts are important. Law enforcement agencies, hospitals, and businesses need to plan extensively. However, effective planning efforts are those that span different organisations and sectors that are guided by a common vision of community resilience in the face of disasters.

Moderation Plans and Mainstreaming Disaster Management into the Development Planning Process

Three committee constitute by Government of India are progressing in the direction of setting up the National Response Plan, National Human Resource and Capacity Development Plan and National Mitigation Plan in respective ministries that have been assigned as nodal offices for different calamities.

The Capacity Development Plan, once it is affirmed and embraced will give the guide to undertaking the capacity building of people occupied with various features of calamity administration and improving the capacity both at the individual and organisation levels.

Working Group on Disaster Management: Planning Commission, GOI, has established a working Group on Disaster Management for giving contributions to the 12th Five-Year Plan (2012–2017) vide no. M-12016/03/2011-PAMD. The terms of reference (particular to working gathering) were as under:

a. To prescribe measures to streamline existing institutional structure on disaster management with a specific goal to maintain a strategic distance from assortment of structures keeping in view the arrangement of the Disaster Management Act, 2005.

b. To review proper implementation towards disaster risk reduction, preparedness and mitigation at centre and state levels and in the private sector.

c. To draw a guide and strategy system to empower open public–private association and network interest in misfortune administration.

d. To evaluate reconciliation of disaster management related worried to be inbuilt focal segment and halfway supported plans/ventures.

e. To recommend programs for capacity building for disaster mitigation at centre, state and region levels with exceptional reference to provincial and urban region.

f. To indentify need zones and ventures alongside monetary assets to be embraced through NDMA. Centre Ministries and State Governments in coordinated way amid the 12th Plan period.

Fortifying the Preparedness Phase

A portion of the illustrative zones and exercises that would decrease the hazard from perils in the readiness stage are condensed as pursues.

Urban Planning and Zoning

There is a need to improve the endeavours for incorporating disaster risk reduction components in settlement arranging and land utilises zoning to mitigate the disaster. Human settlements must be seen from the viewpoint of their helplessness, as well as from the point of view of the calamities that they make or that they exacerbate. There is a need to look at such settlements being created by private manufacturers and engineers, which could expand surge powerlessness in urban and country regions of numerous states. Arranged urban settlements and lodging is the need of the day for disaster risk management that prompts economical advancement, especially in environmentally sensitive region, high-risk prone and high populace thickness pockets.

Construction Laws and Enforcement

Building codes are clung to just in built structures and not in the huge majority of houses over provincial and urban India. The construction laws must be persistently redesigned with the new information and technology.

Lodging Design and Finance

It has been hard to ensure calamity safe innovation at individual level. There is need to discover approaches to support and encourage individual home developers to utilise disaster resilient design, materials and procedures in the development of their homes. Information related with appropriate design and cost effective will be provided for adoption.

Surge Proofing

It is practiced in Bihar, Uttar Pradesh and some parts of India for reducing flood vulnerability. It includes construction of earthen moulds in the house, garden to raise whole properties—the house, the vegetable garden, domesticated animals pen, grain stores, toilets and water wells over the surge level. Empowering such great practices in other flood prone parts of the nation, would go far to reduce the hazard in such zone.

Development of New Financial Instruments

There is need to develop for advancement of new budgetary apparatuses, for example, disastrous hazard financing, risk insurance, calamitous bonds, microfinance and so other.

Risk insurance mechanism would be required for framework, crops and different assets. Making risk insurance obligatory in hazard prone states in the nation should be encouraged.

Agriculture and Aquaculture

These activities ought to be evaluated from the point of view of the flood. Flooding affects villagers and agricultural lands is because of poor drainage system. Mainstreaming disaster risk management should have proper drainage system for development of agricultural practices.

Streets and Infrastructure

Road, electricity supply, railway, communication system should be protect from being destroyed by disaster. Any new framework undertaking should lead a catastrophe affect examination and guarantee that development does not block water stream and cause further or delayed surges. Designing of road should be higher standard to protect from hazard.

Logging Exercises

These activities in the bumpy zones destabilise slants, cause avalanches and increase mudflow and silting in the rivers. The income generated by logging is far lesser than the losses incurred about because of the serious issues of avalanches, silting and environmental caused. There ought to be an arrangement for afforestation in the logging zone previously or soon after the logging.

Build up Disaster Resistance Houses

Housing is on the highest demand among the affected people. Most of the people lived in mud houses that are prone to risk. Because of cyclone and for water logging situation, infrastructure items have been collapsed almost all the shelters including safe sanitation system and this raised the vulnerability for people's livelihood. It was mainly occurred because of insufficient institutional and infrastructural supports for them. It changes their income opportunities. Government and NGOs should take initiative and motivate the people to build up disaster resistant houses. If only one room of every house, capable of accommodating the entire family during the few hours of emergency, may be designed to resist the wind and surge.

Restorative preparedness for recuperation of influenced individual in any sort of debacle is of foremost significance. There ways out immense hole among interest and supply of therapeutic consideration, especially in the zone of injury care. This should be reinforced and limit enlarged at each level from essential to referral level. The limit of specialists and paramedical staff likewise should be fortified and to be outfitted to address the difficulties of post-calamity recuperation.

Major activities ought to be followed with a specific goal to alleviate the disaster:

• Strengthening the State and District Disaster Management Authorities to satisfy their duties as stipulated in the Disaster Management Act, 2005.

• Developing techniques and modalities for ensuring risk reduction through integration of developed projects of all accomplices at national, state and local levels.

• Enhancing the limit with regards to urban risk reduction by tending to arranging capacity building programmes and ensuring reasonable authoritative and administrative instruments to advance safe fabricated condition.

• To fortify the recovery framework, through which the general population influenced by debacles (particularly the most helpless) can get to access for modifying their lives and restoring their occupations.

• To reinforce the knowledge and information sharing stage in disaster management.

INTERNATIONAL STRATEGY FOR DISASTER REDUCTION (ISDR)

The ISDR goes for building disaster resilient communities by advancing expanded attention to the significance of disaster reduction as a necessary part of sustainable development, with the objective of lessening human, social, monetary and ecological misfortunes because of natural as well as man-made disaster.

Tools towards reducing disaster risk for all are as follows.

Increase Awareness to Understand Risk, Vulnerability and Disaster Management

Individuals, governments, non-government organizations, local self-government, United Nation Organization, delegates of civic society and others think about hazard, vulnerability and how to deal with the effects of normal risks, the more mitigation measures will be executed in all divisions of society.

Disaster Risk Reduction Policies and Actions

The more policymakers at all levels concede to disaster reduction approaches and activities, the communities vulnerable against natural disasters will benefit by connected disaster reduction strategies and activities. This requires, to some extent, a grassroot methodology whereby communities in danger are informed and take part in risk management activities.

Empower Interdisciplinary and Intersectoral Organisations, Including the Extension of Risk Reduction Systems

The more organisations involved in disaster risk reduction share data on their research and practices, the more valuable the worldwide group of information and experience will advance. By sharing information and through collective measures we can ensure world that is stronger to the impact of natural hazards.

Enhance Scientific Knowledge about Disaster Reduction

The more we know about the causes and outcomes of natural hazards and man-made disasters on societies, the more we are able to prepared to diminish risk. Bringing researchers, scientists and policymakers together to enable them to contribute and to supplement each other's work.

Mission

- In order to fabricate the resilience of countries and communities to disasters through the usage of the Hygro Framework for Action (HFA), the UNISDR endeavours to catalyse, encourage and assemble the dedication and resources of national, provincial and global partners of the ISDR system.
- The mission of UNISDR is to be a powerful organiser, coordinator and guide all its ISDR partners, globally and regionally, to activate political and financial related duties to disaster risk reduction and Hygro Framework for Action 2005–2015: Building the versatility of countries and commitment to calamities (HFA).
- Develop and manage a robust, multi-partner framework.
- Provide important information and direction.

Function and Responsibilities

The International Strategy for Disaster Reduction (ISDR) was embraced by United Nation members in 2000 and is owned by neighbourhood, national, local and international organizations. UNISDR is driven by a Secretary-General for disaster risk reduction and administered by the Under-Secretary-General Humanitarian Affairs, who serves as the Chair for the wider ISDR system of partnerships. The order of UNISDR is to go about as the point of convergence in the United Nation System for the coordination of disaster reduction and to ensure that disaster risk reduction winds up fundamental part to sound and impartial improvement, natural security and humanitarian activity.

ISDR components: The different mechanisms have been created to accomplish the missions which are specified as follows:

- The biennial Global Platform for Disaster Risk Reduction (GFDRR) goes about as the principle worldwide gathering for proceeded and purposeful accentuation on disaster reduction. Open to all states and the ISDR partners, it basically evaluates the progress made in the execution of the HFA, upgrades consciousness of disaster risk reduction, share experiences and gain from great practices, and identifying gaps and vital activities to quicken national and local implementation.

- It goes about as provincial stages for disaster risk reduction, including ministries gathering, driven by regional intergovernmental associations.

- It acts as a venue for joint work programming among the organisations (FAO, IFRC, ILO, OCHA, UNDP, UNEP, UNESCO, UNICEF, WFP, WHO, WMO and the World Bank). UN organisations give information items and help to provide details regarding improvements (for example, early cautioning, recuperation, training, chance recognisable proof) and so on.

- The Under-Secretary-General for Humanitarian Affairs managed board with its wings (OCHA, UNDG; the World Bank; WMO; UNEP; and IFRC). This Board underpins the seat in giving UN framework wide initiative and high level support for disaster risk reduction.

- Scientific committee, worldwide NGO system for DRR, gender and disaster, media groupings, parliamentarian members and state, which take an interest effectively in ISDR, are called upon in GA and HFA goals to build up multi-partner national stage for disaster risk reduction.

CONCEPT OF DISASTER MANAGEMENT

It may also be termed "a serious disruption of the functioning of society, causing widespread human, material or environmental losses which exceed the ability of the affected society to cope using its own resources" (UN/ISDR, 2004). One of the most difficult concepts in the literature is to arrive at a definition of a disaster. There have been many attempts to define disasters, but all run into the problem of either being too broad or too narrow.

Having a definition of a disaster is extremely important in epidemiology for identifying which events to include or exclude from your analysis. If events are identified with a common definition, then they can also be more easily compared.

In general, most disaster events are defined by the need for external assistance. Perhaps, one reason for this observation is that the disaster relief agencies are often the only organisations with comprehensive and systematic data. There should be some caution applied to data defined in this circumstance. Notably, the decision on which situations require external assistance may differ by country or region. In some situations, it may be a political decision as well. The Centre (CRED) in Brussels uses the following definition. A disaster is a situation or event which overwhelms local capacity, necessitating a request to a national or international level for external assistance, maintains a database of disaster events from 1900 to present. Much of their data is derived from relief groups, including the Red Cross/Red Crescent Agency.

The terms calamity, disasters and hazards are often interchangeably used but they are not synonymous. "Disaster" means a catastrophe or mishap; a grave occurrence affecting any area; arising from nature or man-made cause, or by accident or negligence which results in substantial loss of life of human suffering or damage to, and destruction of property, or damage to or degradation of environment, and in of such a nature or magnitude as to be beyond the copying capacity of the community of the affected area.

Hence, a disaster is said to be take place when it includes two elements, namely hazard and vulnerability.

WHO defines a disaster as any occurrence causing damage, ecological disruption, loss of human lives, deterioration of health and health services on a scale sufficient to warrant any extraordinary interventions from outside the affected community (Lazzari, 1990). WMO defines natural disaster as catastrophic consequence of natural phenomena or a combination of phenomena resulting in injury, loss of life or input in a relatively large scale and some disruption to human activities (WMO, 1989).

Different Phases of Disasters

Both WHO and WMO have looked upon the studies of natural disasters as a continuous sequence essentially for different phases:
a. Pre-disaster phase
b. Impact phase
c. The emergency phase
d. The reconstruction/rehabilitation phase.

Drabek (1986), drawing upon extant studies on human responses to disasters, has proposed a rich inventory and thus has enabled him to put forward a distinct classification system by disaster phase and system level of response. The disaster phases are preparedness (1. planning; 2. warning), response (3. evacuation; 4. emergency), recovery (5. restoration; 6. reconstruction) and mitigation (7. hazard perception; 8. adjustment). The system levels range from the individual to the international.

Disaster Management Cycle—General

Disaster management cycle includes the following stages/phases:
1. Disaster phase
2. Response phase
3. Recovery/rehabilitation phase
4. Risk reduction/mitigation phase
5. Preparedness phase

Figure 3.1 depicts that the process of disaster management involves four phases: Mitigation, preparedness, response, and recovery (Warfield, 2005).

Disaster phase: The phase during which the event of the disaster takes place. This phase is characterised by profound damage to the human society. This damage/loss may be that of human life, loss of property, loss of environment, loss of health or anything else. In this phase, the population is taken by profound shock.

Fig. 3.1: Disaster management cycle

Response phase: This is the period that immediately follows the occurrence of the disaster. In a way, all individuals respond to the disaster, but in their own ways. The ambulances and medical personnel arrive, remove the injured for transportation to medical camps or hospitals and provide first aid and life support. The public also take part in relief work. One can even find injured victims help other injured ones. Almost everyone is willing to help. The needs of the population during this phase are immediate medical help, food—'*roti*', clothing—'*kapda*' and shelter—'*makan*'.

Recovery phase: When the immediate needs of the population are met, when all medical help has arrived and people have settled from the hustle–bustle of the event, they begin to enter the next phase, the recovery phase which is the most significant, in terms of long-term outcome. It is during this time that the victims actually realize the impact of disaster. It is now that they perceive the meaning of the loss that they have suffered. They are often housed in a camp or in some place which is often not their house, along with other victims. During this time, they need intensive mental support so as to facilitate recovery. When the victims have recovered from the trauma both physically and mentally, they realise the need to return back to normal routine. That is, to pre-disaster life. During this phase, they need resources and facilities so as to enable them to return back to their own homes, pursue their occupation, so that they can sustain their life on their own, as the help from the government and other non-governmental organisations are bound to taper in due course. Thus, they are provided with a whole new environment, adequate enough to pursue a normal or at least near-normal life. This is called *rehabilitation*.

Risk reduction phase: During this phase, the population has returned to pre-disaster standards of living. But, they recognise the need for certain measures which may be needed to reduce the extent or impact of damage during the next similar disaster. For example, after an earthquake which caused a lot of damages to improperly built houses, the population begins to rebuild stronger houses and buildings that give away less easily to earthquakes. Or, in the case of tsunami, to avoid housings very close to the shore and the development of a 'green belt'—a thick stretch of trees adjacent to the coast line in order to reduce the impact of the tsunami waves on the land. This process of making the impact less severe is called *mitigation*.

Preparedness phase: This phase involves the development of awareness among the population on the general aspects of disaster and on how to behave in the face of a future disaster. This includes education on warning signs of disasters, methods of safe and successful evacuation and first aid measures. It is worth to note that the time period for each phase may depend on the type and severity of the disaster.

Level of Disasters

For convenience of their management, disasters are classified in four levels.

* *Level '0':* All activities before a disaster strikes come under '0' level. It includes information of disaster management plan, capacity building, construction of disaster resistant buildings especially lifeline buildings, retrofitting weak buildings, trainings of architects, engineers, masons, artisans, etc. mock drills to validate plans and updating identification of available resources and networking between community, village panchayat, taluks, tehsil, resident welfare association, district level training of all stakeholders, identifying the vulnerable groups and catering for their safety, legal framework, spreading general awareness, etc.

* *Level '1':* Disasters which can be managed within the resources available at district level. Examples are road accidents, boat capsizing and low level riots. Domestic fires come under this level.

* *Level '2':* Disasters which can be managed at the state level, under overall guidance of State Disaster Management Authority. There may be movement of personnel and material from one district to another within the state. Examples are low intensity cyclones, floods and even mild intensity earthquakes. In man-made disasters, rail accidents, multiple road accidents, urban fires, etc. fall under Level '2' disasters.

* *Level '3':* All disasters which are outside the coping capacity of the state and require intervention from the central government comes under Level '3' disasters. Bhuj earthquake, Tsunami 2004 and Orissa Super Cyclones 1999 come under Level '3' disasters.

Approaches to Disaster

Alexander (1993) identified six schools of thought on natural hazards and disaster studies: The geographical approach, the anthropological approach, the sociological approach, the development studies approach, the disaster medicine approach and the technical approach (Fig. 3.2).

The geographical approach (pioneered by Barrows, 1923 and White, 1945) deals with the human ecological adaptation to the environment with special emphasis on the 'spatio-temporal' distribution of hazard impacts, vulnerability and people's choice adjustment to natural hazards. Social science methods are widely used in this approach.

Disaster Management

The term 'disaster' may be defined as an unexpected happening causing a huge loss of

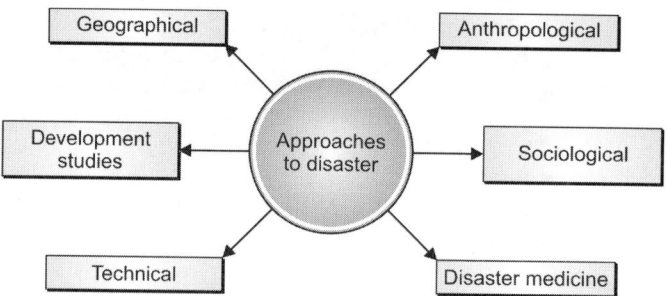

Fig. 3.2: Approaches to disaster research (Alexander, 1993)

life and property. It may be both normal as well as man-made, in the sense that in some parts of the country, floods are coming due to construction of check dams and embankments on the rivers. Disaster may come and ruin the life and property of the people in the forms of earthquakes, cyclones, floods, drought, fire, etc. According to one estimate, about 5200 disasters took place in 178 countries of the world during 1963–1992. Out of these, 787 disasters took place in Asia between 1963 and 1992 causing more than 100 deaths in each case. The major disaster may be defined as those causing at least one of the three consequences which are as follows:

a. Causing damage to more than 1% of Gross National Product (GNP); or

b. Affecting more than 1 % of total population of an area; or

c. Causing deaths of more than 100 persons at a time and place.

According to another estimate during 1950–1992 about 5 lakh deaths took place out of which 1% in developed countries and 99% in developing countries. Before 1950 the situation was quite different and at that time 30% of total deaths took place in developed countries and 70% in developing countries. These figures indicate the gravity of the situation and seriousness of the problem.

A collective term encompassing all aspects of planning for and responding to disasters, including both pre- and post-disaster activities. It refers to the management of both the risks and consequences of disasters.

The terms 'disaster mitigation' and 'management' are used many times inter-changeably. However, there is a thin line of differentiation between mitigation and management. In simplest terms disaster mitigation and management both include efforts and practices that seek to reduce the ill effects of disasters. However, disaster management is a more holistic term, which encompasses a larger framework of issues in comparison to mitigation efforts. The aim of both mitigation and management strategies is to reduce the losses in the event of a future occurrence of a hazard. The primary aim is to reduce the risk of death and injury to the population. Secondary aim includes reducing damage and economic losses inflicted on the community as a whole. The objectives are likely to include encouragement for people to protect themselves as far as possible. Some disasters can be avoided. This can be done through removal of hazard from the potential area. In realistic terms, however, not all disasters can be totally prevented. Then the key is to mitigate the damage.

Disaster management includes the development of disaster recovery plans (for minimising the risk of disasters and for handling them when they do occur) and the implementation of such plans. Disaster management usually refers to the management of natural catastrophes such as fire, flooding, or earthquakes. Related techniques include crisis management, contingency management, and risk management. In a holistic term it includes the following:

• It is more than just response and relief (i.e. it assumes a more proactive approach).

• It is a systematic process (i.e., based on the key management principles of *planning, organising,* and *leading* which include *coordinating* and *controlling*).

- Aims to reduce the negative impact or consequences of adverse events (i.e. disasters cannot always be prevented, but the adverse effects can be minimised).
- It is a system with many components.

Disaster/emergency management is the discipline of dealing with and avoiding risks. It involves preparing for a disaster before it happens, disaster response (e.g. emergency evacuation, quarantine, mass deconta-mination, etc.), as well as supporting, and rebuilding society after natural or human-made disasters have occurred.

In general, any emergency management is the continuous process by which all indi-viduals, groups, and communities manage hazards in an effort to avoid or ameliorate the impact of disasters resulting from the hazards. Actions taken depend in part on perceptions of risk of those exposed. Effective emergency management relies on thorough integration of emergency plans at all levels of government and non-government involve-ment. Activities at each level (individual, group, community) affect the other levels. It is common to place the responsibility for governmental emergency management with the institutions for civel defence or within the conventional structure of the emergency services. In the private sector, emergency management is sometimes referred to as business continuity planning.

Other terms used for disaster management include:

- *Emergency management* which has repla-ced *Civil defence* can be seen as a more general intent to protect the civilian popu-lation in times of peace as well as in times of war.
- *Civil protection* is widely used within the European Union and refers to government approved systems and resources whose task is to protect the civilian population, primarily in the event of natural and human-made disasters.

- *Crisis management:* It is the term widely used in EU countries and it emphasizes the political and security dimension rather than measures to satisfy the immediate needs of the civilian population.
- *Disaster risk reduction:* An academic trend is towards using the term is growing, particularly for emergency management in a development management context. This focuses on the mitigation and preparedness aspects of the emergency cycle.

Mitigation: It is the collective term used to encompass all actions taken prior to the occurrence of a disaster (pre-disaster measures) including preparedness and long-term risk reduction measures (mitigation and preparedness sometimes used alternatively).

Disaster Preparedness

Social scientists, emergency managers, and public policymakers generally organise both research and guidance around four phases of disaster loss reduction: Mitigation, prepared-ness, response, and recovery. According to report of the National Research Council (NRC, 2006), the core topics of hazards and disaster research include: Hazards research, which focuses on pre-disaster hazard, vulnerability analysis and mitigation; and disaster research, which focuses on post-disaster emergency response and recovery. Preparedness inter-sects with both of these two areas, serving as a temporal connector between the pre-impact and post-impact phases of a disaster event. Preparedness is typically understood as consisting of measures that enable different units of analysis—individuals, households, organisations, communities, and societies—to respond effectively and recover more quickly when disasters strike. Preparedness efforts also aim at ensuring that the resources necessary for responding effectively in the event of a disaster are in place, and that those faced with having to respond know how to use those resources. The activities that are commonly associated with disaster pre-paredness include developing planning processes to ensure readiness; formulating disaster plans; stockpiling resources necessary for effective response; and developing skills and competencies to ensure effective performance of disaster-related tasks.

The concept of disaster preparedness encompasses measures aimed at enhancing life safety when a disaster occurs such as protective actions during an earthquake, hazardous materials spill, or terrorist attack. It also includes actions designed to enhance the ability to undertake emergency actions in order to protect property and contain disaster damage and disruption, as well as the ability to engage in post-disaster restoration and early recovery activities. The aims of disaster preparedness are minimise to adverse effects of a hazards through effective precautionary actions, and to ensure timely, appropriate and efficient organisation and delivery of emergency response following the impact of a disaster.

Preparedness is commonly viewed as consisting of activities aimed at improving response activities and coping capabilities. However, emphasis is increasingly being placed on *recovery preparedness*—that is, on planning not only in order to respond effectively during and immediately after disasters but also in order to successfully navigate challenges associated with short- and long-term recovery. FEMA (Federal Emergency Management Agency) defines preparedness as: The leadership, training, readiness and exercise support, and technical and financial assistance to strengthen citizens, communities, state, local, and tribal governments, and professional emergency workers as they prepare for disasters, mitigate the effects of disasters, respond to community needs after a disaster, and launch effective recovery efforts. The capability assessment for readiness (CAR), which was developed by FEMA and the National Emergency Management Association (NEMA), identifies thirteen elements that should be addresses by states in their preparedness efforts. Those elements are:

• Laws and authorities
• Hazard identification and risk assessment
• Hazard mitigation

• Resource management
• Direction, control, and coordination
• Communications and warning
• Operations and procedures
• Logistics and facilities
• Training
• Exercises, evaluations, and corrective actions
• Crisis communications, public education, and information
• Finance and administration.

Mitigation and preparedness are sometimes conflated with one another (as they are in the list above), in part because they are intertwined in practice. Indeed, definitions contained in key resource documents reviewed for this project illustrate this conceptual blurring. For instance, the National Fire Protection Association (NFPA) defines preparedness as: Activities, programs, and systems developed and implemented prior to a disaster/emergency that are used to support and enhance mitigation of, response to, and recovery from disaster/emergencies (NFPA, 2004).

The Role of Disaster Management

Ultimately, disaster management aims to reduce the impact of disasters. The ways of achieving this have varied and evolved over time. The earliest and still predominate approach is for agencies to provide relief to those affected once a disaster has happened. Rescue assistance, medical support, food and water supply are vital for saving lives which prevent further harm. However, responding to a disaster can only do so much, and a level of loss is almost inevitable before a rescue operation can even arrive. As well as wasting precious time, relying on external support is not desirable for communities at risk of a disaster, particularly as they often have a capacity to deal with a disaster already.

More recent approaches such as that followed by Practical Action, take a more holistic view and seek to reduce the risk of a disaster. Rather than waiting to respond, disaster management programmes plan for the whole disaster process including a range of activities at different stages of disaster management. The strategies include risk reduction such as hazard, exposure and sensitivity reduction, impact reduction, and capacity building for resilience addressing not only the impacts but also the factors that turn a hazard into a disaster. Poverty is well understood to significantly increase the disaster impact, a fact that is observable in nearly all events around the world as it is the poorest that are the most affected each time. This leads to the view that disasters are a symptom of incomplete, inappropriate or inequitable development. As a development organisation that specialises in working with communities to develop technologies to alleviate poverty, Practical Action aims to break the cycle that makes people in poverty particularly susceptible to disasters.

NATIONAL DISASTER MANAGEMENT FRAMEWORK

The institutional structure for disaster management in India is in a condition of change. The new set up, following the execution of the act, is advancing; while the past structure also continues with. Subse-

quently, the two structures exist together at present. The National Disaster Management Authority has been built up at the centre, and the SDMA at state and DDMA at district level are step by step being formalised. Besides these the National Crisis Management Committee, was in the earlier set up, additionally works at the centre. The nodal ministries, as distinguished for various disaster types of function under the direction of the Ministry of Home Affairs. This influences the stakeholders to collaborate at various levels inside the disaster management framework which was set up by Government of India. Inside this set up, two distinct features for disaster management might be noted. Figure 3.3 describes that the structure is hierarchical and worked at four levels— Centre, state, district and local. In both the set ups—one that existed prior to implementation of the act, and other that is being formalised post-implementation of the act. Each level guides the exercises and basic leadership at the following level in hierarchy. Besides, it is a multi-partner set up, i.e. the structure draws association of different important services, government departments and regulatory bodies.

The NDMA, as the apex body for Disaster Management is headed by the Prime Minister.

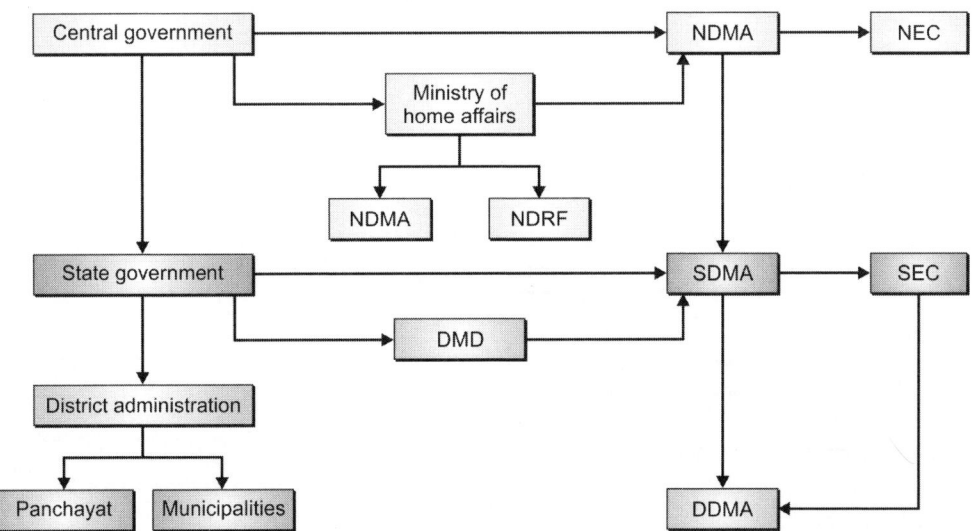

Fig. 3.3: National Disaster Management Framework

The state authorities lay down the guidelines for departments of the state and the district falling in their respective jurisdictions. Similarly, district authorities direct the civil administration, departments of local authorities such as the municipalities, police department and civil administration. The State Executive Committee are responsible for execution of the tasks envisaged by the authorities. The structure thus discuss is summarised in Fig. 3.4.

FINANCIAL ARRANGEMENTS

The arrangement and the financing policy for arrangement of assistance help to those victims by common disasters are clearly set down. Finance Commission, Government of India reviewed this in every year. The Finance Commission makes proposal with respect to revenue gained from tax and non-tax between centre and state government and furthermore in regards to provision of relief assistance and

share of expenditure between central and state government. As per the recommendation of Eleventh Finance Commission, a Calamity Relief Fund (CRF) has been set up in each state. Based on the expenditure of relief and rehabilitation over the past ten years, the size of the CRF will be fixed by Finance Commission. The share between central and state government of the CRF is 75: 25 in each state. The CRF is used for response and relief as and when required. By and large, relief assistance are set around a national board of trustees with members of states. Each states can have state-specific norms to be prescribed by state level board under the Secretary. Where the catastrophe is of such extent that the fund in the CRF will not be adequate for arrangement of alleviation, the state looks for help from the National Calamity Contingency Fund (NCCF). At the point when such demands are received from particular state, the demand will assessed by a group from the central government placed

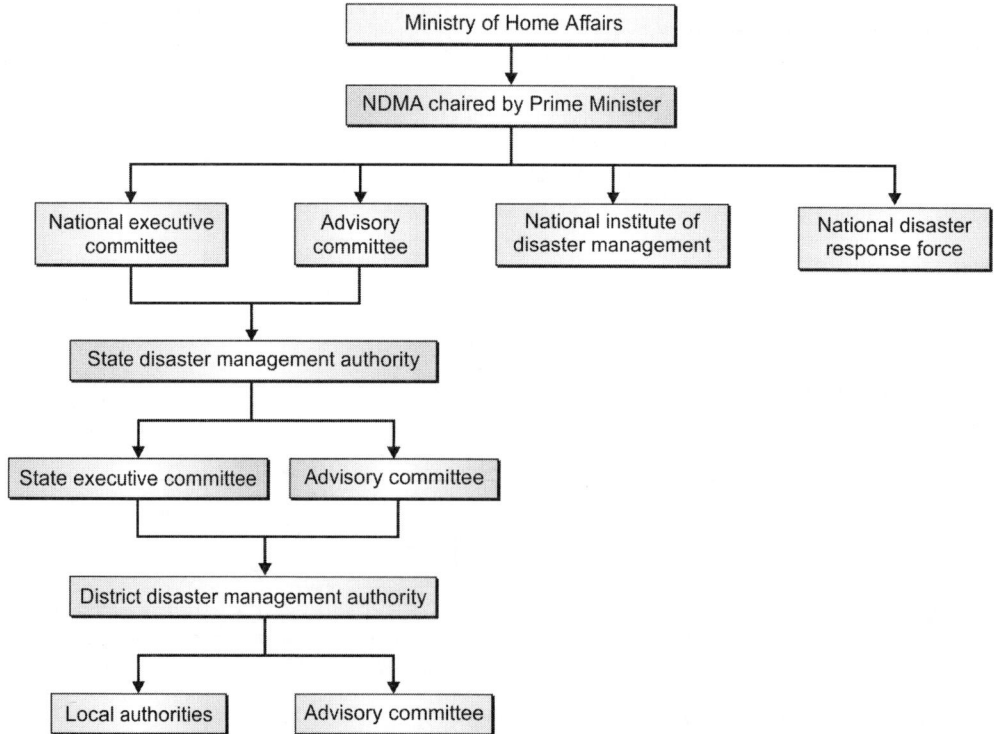

Fig. 3.4: Flowchart of Administrative set up of Disaster Management Committee in India

before an Inter-Ministerial Group chaired by the Home Secretary. High Level Committee led by the Home Minister with the Agricultural Minister, Finance Minister and the Deputy Chairman, Planning Commission take decision in regards to the release of fund from the NCCF dependent on the report of the Central team and the proposals of the IMG consequently. To sum things up, the institutional arrangement for relief and rehabilitation are settled and have ended up being powerful and successful.

ROLE OF NGOs

NGOs are organizations that are nonprofit making, voluntary and independent of government, engaged in activities concerning various societal and developmental issues. The role of NGOs during a disaster is to have quick response and to try and save as many lives as it can with the given funds.

The specific role of NGOs in respect to disaster management is:

Pre-disaster

- Training and capacity building of NGO staffs and task forces;
- Set up of information channel to the village and district;
- Advocacy and planning; and
- Regular contact with block control room.

During disaster

- Activate of channel of warning dissemination to reach the target groups;
- Help block administration for wide dissemination of warning;
- Immediate rescue and first aid, including psychological aid, supply of food, water, medicines, and other immediate need materials;
- Ensuring sanitation and hygiene; and
- Damage assessment.

Post-disaster

- Technical and material aid in reconstruction;
- Assistance in seeking financial aid; and
- Monitoring.

Education

NGOs provide training and education to the children and people in disaster affected areas. NGOs jointly with the education department will undertake different programmes for children on school safety, vulnerability assessment of schools and assist in developing school disaster management. Nowadays, disaster management should included a main subject in school curriculum development for educational institutions.

Health Including Reproductive Health (HIV/AIDS)

NGOs, providing in the health and sanitation facility in the disaster affected area in consultation with health department and district administration. NGOs planed the details on the various needs with regard to emergency health and gender issues in the local level, availability of health professionals, medicine supply, preposition of dignity/hygiene kits and the implementation plan. They may also develop health volunteers within the community and provide training to the volunteers on basic health care services including first aid and ensure practice through periodic mock drills.

Shelter

Losses due to destruction of houses and infrastructure took places in just during and after the disasters. People are shifted to the shelters. NGOs may render technical support in consultation with the administration for arranging disaster resilient shelters, infrastructures and so onto reduce the impact of disaster and lesser losses.

Employment (Agriculture/Animals)

People in India are self-sufficient and largely dependent on agriculture to generate their livelihood. After disaster, people loss their crops and as well as agricultural land for cultivating of crops. Therefore NGOs assist the administration in technical support for generation of alternate livelihood opportunities.

Water and Sanitation

NGOs provided sufficient quantity of safe drinking water and foods to the villagers. NGOs would also arrange the sanitation facilities such as assess to toilet, vector issues and strong waste management, sewerage and seepage.

Protection

NGOs job in setting up protection mechanisms prevent vulnerable peoples affected by the disasters. NGOs will assist the local self-government in forecasting weather information time to time through public address system.

Environmental Conservation

NGOs will provide technical support with the government administration in awareness generation, sensitisation and information dissemination for conservation of natural resources and climate change adaptation.

Disability

Prior to any disaster, the NGOs would prepared the villagers/victims in identification of vulnerable groups, warning mechanisms, providing food grains, arranging shelters, identification of resources and assets, household preparation, capacity building and developing a disaster management plan, etc. This needs a sustained effort on the part of the community.

COMMUNITY-BASED DISASTER MANAGEMENT

Top-down management approaches were simply unsuccessful in addressing the needs of vulnerable communities. A better understanding of disasters and losses reveals that the increase in disaster occurrence and disaster related loss is due to the exponential increase in occurrence of small and medium scale disasters. As a result many feel it is important to adopt a new strategy, which directly involves vulnerable people themselves in planning and implementation of mitigation measures. This bottom-up approach has received wide acceptance because communities are considered the best judges of their own vulnerability and can make the best decisions regarding their own well-being.

The aim of CBDM is to reduce vulnerabilities and strengthen people's capacity to cope with hazards. A thorough assessment of a community's exposure to hazards and an analysis of their specific vulnerabilities and capacities is the basis for activities, projects and programs that can reduce disaster risks. Because a community is involved in the whole process, their felt and real needs, as well as inherent resources, are considered. It is therefore more likely that appropriate interventions will be used.

People's participation concerns both processes and content. The community should be able to gain directly from improved disaster risk management. This, in turn, will contribute to a progression towards safer conditions, security of livelihood and sustainable development. This underlines the point that the community is not only the primary actor but also the beneficiary of the risk reduction and development process. Some authors differentiate between community participation and involvement. Community participation is generally taken to mean that a given community takes responsibility at all stages of a program, including planning and implementation. Community involvement refers to a 'less than' ideal situation, where the community is asked to participate in a program that has already been designed by someone else.

Implementation of CBDM points to the following essential features:

i. *The community's central role.* The focus of attention in long-term and short-term disaster management must be the local community.

ii. *Disaster risk or vulnerability reduction as the foundation of CBDM.* The primary content of disaster management activities revolve around reducing vulnerable conditions and the root causes of vulnerability. The primary strategy of vulnerability reduction is to increase a community's capacities, resources and coping strategies.

iii. *Linkage to the development process.* Disasters are viewed as unmanaged development risks and unresolved problems of the development process. CBDM should lead to a general improvement of the quality of life of the vast majority of the poor people and of the natural environment. CBDM contributes to people's empowerment—to possess physical safety; to have more access and control of resources; to participate in decision making that affects their lives; to enjoy the benefits of a healthy environment.

iv. *Community as a key resource in disaster risk reduction.* The community is the key actor as well as the primary beneficiary of disaster risk reduction. Within the community, priority attention is given to the conditions of the most vulnerable as well as to their mobilization in disaster risk reduction. The community participates in the whole process of disaster risk management from situational analysis to planning to implementation.

v. *Application of multi-sectoral and multi-disciplinary approaches.* CBDM brings together the multitude of community stakeholders for disaster risk reduction to expand its resource base. The local community level links up with the intermediate and national and even up to the international level to address the complexity of vulnerability issues. A wide range of approaches to disaster risk reduction is employed.

vi. *CBDM as an involving and dynamic framework.* Lessons learned from practice continue to build into the theory of CBDM. The sharing of experiences, methodologies and tools by communities and CBDM practitioners continues to enrich practice.

Before implementing CBDM it is important to know who in the community should be involved. The most vulnerable are the primary actors in a community. The focus should be at the household level. As all individuals, houses, organisations and services stand a chance of being affected, they should all be involved. But before working on disaster risk reduction, differing perceptions, interests, and methodologies have to be recognised and a broad consensus on targets, strategies and methodologies have to be reached.

ROLE OF MEDIA

Media play a crucial role during disaster particularly in preparedness phase. It can be used as a useful media for continuously disseminating information during disaster by spreading awareness, transmitting messages through helpline, emergency contact number, heath care, updated information about the disaster and needs of the people.

Disasters are source of news and it captured the attention of the world about the disaster related issues. The areas where media can contribute are as follows (Dave, 2009):

1. Aid prioritisation of disaster risk issue: The media can impact the legislature to organise calamity chance issues, consequently ensuring that "self serving" political interests are not underscored to the detriment of the more extensive populace.

2. Early warning information: The media can help in telecasting messages about the early warning, proving information on risk and existing technologies that can help the improvement of valuable ideas and frameworks. The effective instruments for proving information are radio, television and cable services.

3. Increase worldwide gifts: The media can trigger donation from the global network consequent to the event of national calamities, and also push the administration to increase the budgetary allocation for disaster risk reduction program.

4. Improved coordination of risk assessment activities: The media can enhance the coordination of risk assessment between policymakers and donors. This coordination will result increased availability of local resources therefore save lives and population as a whole.

Effect of Media

The media is normally the first to providing information in the occasion as an official calamity. At first, they informed the general population about it and therefore create awareness. After that people knew how it is being managed.

1. The media are considered as important sources of information in the local level for providing continuous information about the disasters.
2. Continuously forecasted of information about the occurrences and post-disaster can help basic leadership and reaction quickly after a calamity, therefore saving lives and property.
3. The media is a significant resource by proving important information about the safety, availability of transport and communication system, etc. and so on.
4. It can also be used as providing information on location of distribution of relief materials, health camp, drinking water supply, etc. organised by government in a particular area for people in general.

Organisations with the Media

Media has a task to carry out all phases of calamities. Before disaster, the media continuously disseminate early warning to the vulnerable population.

After a catastrophe has struck, news media can give powerful resources for information about the progress of disaster, how much it affected the population, property and thus helping the administrators to guide and safeguard endeavours to survivors.

Media can assume an exceptionally compelling job in instructing people in general about debacles; cautioning of dangers;

assembling and transmitting information about affected regions; alarming government authorities, help associations, and the general population to particular needs; and encouraging dialogues about calamity readiness and reaction.

NATIONAL DISASTER MANAGEMENT AUTHORITY (NDMA)

At the national level, the Ministry of Home Affairs is the nodal agency for all issues concerning disaster management. The Central Relief Commissioner (CRC) in the Ministry of Home Affairs undertakings is the nodal officer to organise help tasks for catastrophic events. The CRC gets information from India Meteorological Division (IMD) about the warning/forecasting of natural disasters. The Department/Offices/Associations concerned about the disaster management include: India Meteorological Department, Central Water Commission, Ministry of Home Affairs, Ministry of Finance, Ministry of Rural Development, Department of Agriculture and Cooperation, Ministry of Railway, Ministry of Information and Broadcasting, Ministry of Petroleum, Branch of Farming and Participation. Service of Intensity, Division of Common Supplies, Planning Commission, Ministry of Social Justice, Department of Women and Child Development, Ministry of Environment and Forest, Department of Food. Every Department/Ministry/Association chooses their nodal officer to the Crisis Management Group headed by Central Relief Commissioner. The nodal officer is in responsible for preparing Disaster Management Plan.

Mandate of NDMA

The NDMA laid down policies and guidelines for disaster management which would be followed by different sectors, Ministry, Departments, state government in taking measures for disaster mitigation. The guidelines and policies are as follows:

a. Lay down approaches on disaster management;

b. Approve the national plan;

c. Approve plan prepared by different ministries, departments as per the national plan;

d. Set down rules to be trailed by the state departments in drawing up the state plan;

e. Set down rules to be trailed by the different sectors of the Legislature of India to integrate the measures for avoidance of calamity;

f. Arrange the requirement and execution of the approach and plan for disaster management;

g. Recommend provision of fund for disaster mitigation;

h. Provide support to different countries affected by disaster;

i. Take such different measures for the preparedness, mitigation, capacity building for dealing with the negative effect of disasters; and

j. Set down polices and guidelines for functioning of the National Institute of Disaster Management.

National Executive Committee (NEC)

A National Executive Committee is constituted to help the National Expert in the execution of its capacities. NEC comprises Home Secretary as its Director, ex-officio, with different Secretaries to the Legislature of India in the Services or Offices having regulatory control of the farming, nuclear vitality, guard, drinking water supply, condition and timberland, fund (use), well-being, control, provincial advancement science and innovation, space, media transmission, urban improvement, water assets. The Head of Incorporated Resistance Staff of the Head of Staff Advisory group, ex-officio, is additionally its individuals.

NEC may as and when it considers essential comprise at least one sub-committee for the proficient release of its capacities. NEC has been given the responsibility to go about as the planning and coordinating body for disaster management, to set up a national plan.

STATE DISASTER MANAGEMENT AUTHORITY (SDMA)

The Disaster Management Act, 2005 provides constitution of SDMAs and DDMAs in every one of the states and UTs. Except one state Gujarat and one UT, i.e. Daman and Diu, all the rest have established SDMAs under the DM Act, 2005. Gujarat has established its SDMA under its Gujarat State Disaster Management Act, 2003. Daman and Diu has additionally settled SDMAs preceding sanctioning of DM Act, 2005.

With regards to bureaucratic set up of India, the response to natural disaster is basically that of the concerned state government. However, central government provides financial, physical and assistance to natural disasters. The factors responsible for response to state governments are as follows:

i. Gravity of a natural disaster

ii. Relief operations necessary

iii. Central assistance for financial help.

Each and every state has Relief Commissioners, under the department of disaster management, who are responsible for the relief measures in the wake of catastrophic events. He supervises and controls the relief operation through District Collector at district level.

State Crisis Management Group (SCMG)

Every state has is a State Crisis Management Group (SCMG) under the Chairmanship of Chief Secretary. This consists of senior officers from the department of revenue, home, power supply, panchayat, water supply, agriculture, rural development and health, forest, public work.

Government of India directs the SCMG to guide and formulate action plan for dealing with disasters. It is the duty of every state to establish a control room at the state headquarters as soon as the disaster occurred. Not only forecasted information about the early warning, SCMG would also contact with the all departments.

State Executive Committee (SEC)

The role of SEC is to planning, coordinating and monitoring the national plan, national policy and state plan. SEC chaired by the Chief Secretary of the concern state with four different secretaries of such divisions as the state government may think fit.

DISTRICT DISASTER MANAGEMENT AUTHORITY (DDMA)

The District Disaster Management Department formulates and carries out the plan under the overall guidance of the District Magistrate. However, it is impossible for the Relief Department alone to formulate the plan without the cooperations of the various lines departments that are concerned with disaster management. Prominent among such departments are Irrigation and Waterways, Public Health Engineering, Health, Animal Husbandry, Agriculture, Food and Supplies, Public Works Department, Telecommunication, Electricity, etc. Last but not the least the Police Department has a very major role in carrying out the plan. This Committee takes stock of the situation, monitors routine preparedness, suggests improving response mechanism and develops a document for disaster management in the district. It has been decided that the committee shall sit at least twice in a year for the above purpose.

ROLE OF LOCAL ADMINISTRATION

The PRI is a statutory body elected by the local people through a well-defined democratic process with specific responsibilities and duties. The elected members are accountable to the people of the ward, rural community, block and the district. Keeping the above in view, the PRI, the representative body of the people, is the most appropriate institution from village to the district level in view of its proximity, universal coverage and enlisting people's participation on an institutionalised basis. Their close involvement will go a long way in getting people prepared for countering natural disasters as well as involve them in all possible preventive and protective activities so that the impact of the disasters are mitigated and the people are able to save their lives and property. The PRIs can act as catalysts to social mobilisation process and tap the traditional wisdom of the local communities to complement the modern practices in disaster mitigation efforts. Besides PRIs will also provide a base for integration of various concerns of the community with that of the NGOs and community-based organizations (CBOs) which are engaged in various developmental activities at the grassroots level.

Need for Involving the PRI Bodies

In general, if the local bodies like Panchayats are not consulted for preparedness—planning, relief and rehabilitation work, it leads to absence of transparency and accountability in the mitigation efforts. The whole approach towards rehabilitation work may end up being, top down in nature. As the relief and restoration efforts involve investment of hundreds and thousands crores of rupees, there should be satisfaction of having utilised them properly and efficiently. Activities like distributing immediate relief in the form of money, food grains, medical care, cloths, tents, vessels drinking water and other necessities, activities of restoration, rehabilitation and reconstruction efforts of damaged villages and towns can be implemented better with the involvement of local bodies. There is a view that local bodies like Panchayats should be encouraged and empowered to manage the local affairs with the available local resources. The elected leaders and officials of Panchayats should be trained to develop capabilities to handle crisis situation in preparedness, warning, rescue, relief, medical assistance, damage assessment, counseling, water and sanitation and rehabilitation operations. It is felt that in biggest disasters the role assigned to Panchayats was meager in handling the problems of various types at the grassroots level.

The 73rd Constitution Amendment (1992) heralded a new phase in the country's quest for a democratic decentralised set up; more so, in matters pertaining to devolution of powers, functions, functionaries and finances. One of the objectives of Panchayati Raj (PR) is to promote popular participation through an institutional framework. The articles 243(G) of the Constitution visualises Panchayats as institutions of self-government. It subjects to extent of devolution and powers and functions to the will of the state legislatures, it also outlines the role of Panchayats in respect of development, planning and implementation of programs of economic development and social justice. A comprehensive list covering 29 subjects which are mostly related to development has also been provided in the Eleventh Schedule to the constitution. The success of this depends upon adequate devolution of powers, functions, personnel and finances on these bodies, which is yet to make significant progress. Mostly the disaster activities of restoration rehabilitation and reconstruction fall within the ambit of these development activities. Hence, there is an imperative need to involve local bodies in disaster management (Goel, 2006).

The PRI members can play a role of leadership in Disaster Management at all stages. Right from the preparatory stage up to the handling of the long-term development activities for risk reduction, PRI can lead in several ways. A broad outline may include activities in Table 3.1.

Table 3.1: Role of PRI according to GOI-UNDP, 2009

Sl. No.	Phase		
	Pre-disaster	*During disaster*	*Post-disaster*
1.	Organising awareness campaign and promoting community education on disaster preparedness	Arranging emergency communication through available resources	Damage assessment particularly assisting in identifying victims for compensation and its distribution
2.	Articulation of community need for developing preparedness plan through community involvement and Panchayat ownership	Evacuation to temporary shelter and running relief camps	Formulating rehabilitation and reconstruction plan of houses and other local infrastructures
3.	Identifying the resource gaps both physical and manpower and replenish the same through capacity building	Supplementing rescue and relief efforts in coordinating different agencies	Enforce minimum specification for safe reconstruction
4.	Establishing synergy with local agencies including NGOs/CBOs	Monitoring of relief distribution	Supervise and monitor long-term reconstruction and mitigation projects
5.	Activating the DM Plans with the participation of the community	Safe disposal of carcass and arranging safe drinking water and sanitation	
6.	Encouraging people to insure assets and livestock		
7.	Formation of Task Forces and their capacity building		
8.	Establishing convergence with local institutional structures created for implementing education, health, livelihood, social justice and so on		

ARMED FORCE IN DISASTER RESPONSE

National Disaster Response Force (NDRF)

Role of NDRF: The primary assignment of NDRF is as follows:

- Removal of debris.
- Moving victims to a safer places.
- Provide moral support to the victims.
- Assistance in distributing the relief material.
- Coordinate with sister organisations.
- Capacity building.
- Giving help to outside nations, whenever inquired.

Other Activities

NDRF is occupied with following different exercises.

- Conduct familiarisation exercises with the victims during the disasters.
- Coordinate with other stakeholders.
- To direct community awareness program for capacity building.
- To organise exhibition.
- To organise training for development/ enhancement of skill of the NDRF.
- To prepare State Disaster Response Force (SDRF), community and NGO's in calamity management.

CIVIL DEFENCE

Civil defence is a community-based voluntary organisation and cannot only to rescue and rehabilitation measures but also play important role in public awareness and capacity building to face any disasters.

The Civil Defence policy confined to making all states and UTs conscious of the need for civil protection measures and requesting Civel Defence to keep prepared common civil protection plan in urban areas and towns under the Emergency Relief Operation [ERO].

According to DM Act, 2005, it is required for NDMA to ensure civel defence preparedness for disaster management. The roles of Civil Defence in respect of disaster management are as follows.

Pre-disaster

- Creating awareness about the various kinds of disasters.
- Educating/training the general population for response to any disaster.
- Transmitting messages related with disasters to the people.
- Holding consistent mock drill, activities and practice of civel defence activities.
- Prepare and distribute publicly material, brochures about civel defence.
- Provide lecture, demonstration, showing films.
- Organise camp to provide basic training skill for disaster management.

During and After Disaster

- After getting warning information, prepared measures taken immediately after getting warning information.
- Helping in evacuation of victims to a safer places.
- Providing/arranging first aid treatment.
- Help in search and rescue operation.
- Provide information about missing, injured people and facilities available to the affected areas.
- Participating in distribution of relief materials to the people.
- Assisting police in ensuring smooth transport facilities in the affected areas.
- Helping the local government in assessing loss of lives and property.
- Conducting disaster awareness training in the districts.

DISASTER RESPONSE

Disaster response refers to decision and actions taken in order to tactic, strategic and operational goals characterised by disaster responders. Disaster response comprises various components, for instance; warning, rescue, providing assistance, assessing loss, distributing relief materials, shifting vulnerable communities to safer places, arranging shelter, dissemination of information, arranging communication facilities, immediate restoration of infrastructure, etc. for immediate respond to disasters to protect life and property. The point of crisis reaction is to give quick help to look after life, enhance well-being and bolster the confidence of the influenced populace. The main objective is to putting people safe, mitigate disasters and meeting the essential needs of the general population during and after disaster. The government, municipalities or panchayats are the primarily responsible to respond the disaster in that particular area, locality where the disaster has occurred. Though NGOs are also responsible to mitigate and meeting the basic needs of the affected population.

Objectives for disaster responders are:

- Saving and securing human life.
- Mitigate the disaster.
- Providing information about early warning.
- Protection measures for health.
- Restoring relief materials.
- Arranging drinking water facilities.
- Arranging emergency communication through available resources.
- Facilitating early recovery by providing help in physical, economic, mental, social.
- Assess damage and loss.
- Organised training and education.
- Identifying and taking action to lesson learned.
- Facilitating the recuperation of the community.
- Evaluating the response and recovery activities.

In India, the National Disaster Management Authority is responsible for planning and mitigating alleviating impacts of disastrous events and anticipating and avoiding man-made disasters. It coordinates with different line department and government agencies to mitigate and emergencies (NDMA, 2014). The National Disaster Response Force is an inter-governmental disaster reaction organisation that specialises in search, rescue and rehabilitation (NDRF, 2014).

ROLE OF POLICE

During disaster, police not only keep on protecting the community from conceivable plundering, destruction of property, and burglary that may happen, they likewise must be set up to evacuate people, render propelled life-saving methods, and keep dispensing sites secure. In addition to these the police are responsible for distribution of food, cloths, water, etc. to the people who had been affected by disasters.

Job of police
- Crowd control.
- Looting.
- Traffic control.
- Search and protect.
- Coordination with government and other agencies.

Job of firefighters
- Control fires that threaten basic framework.
- Deal with perilous materials.
- Provide first aid medical care.
- Transport and communicate victims.
- Search and rescue.
- Coordination with government and non-government agencies.

Job of EMS (Emergency Medical Service)
- Response and recovery.
- Primary medical care.
- Transfer vulnerable people to hospital.
- Search and protect.

- Coordination with government and non-government agencies.

MODELS OF DISASTER MANAGEMENT

A Comprehensive Conceptual Model for Disaster Management

Kelly (1998), states that, there are four main reasons why a disaster model can be useful. These are as follows:

1. A model can simplify complex events by helping to distinguish between critical elements. Its usefulness is more significant when responding to disasters with severe time constraints.
2. Comparing actual conditions with a theoretical model can lead to a better understanding of the current situation and can thus facilitate the planning process and the comprehensive completion of disaster management plans.
3. The availability of a disaster management model is an essential element in quantifying disaster events.
4. A documented disaster management model helps establish a common base of understanding for all involved. It also allows for better integration of the relief and recovery efforts.

Therefore, based on the above, it can be argued that a well-defined and clear model is highly beneficial in the management of disasters because it facilitates the securing of support for disaster management efforts. Hence, disaster management needs a formal system, or a model, to manage and possibly reduce the negative consequences of a disaster.

Based on a survey of relevant literature, the present researcher has separated different disaster management models into the following main categories: Logical, integrated, causes and others. Existing disaster management models, fit into one of these categories, are shown in Fig. 3.5. Logical models, provide a simple definition of disaster stages and emphasize the basic events and actions which constitute a disaster. Integrated models characterise the phases of a disaster by the evolution of functions such as strategic planning and monitoring. In these models, modules are linked as events and actions. The cause category, which is not based on the idea of defining stages in a disaster, suggests some underlying causes of disasters. The last category, describes miscellaneous models.

The traditional process of disaster management consists of two phases: (1) Pre-disaster risk-reduction and (2) post-disaster recovery phase. The first consists of activities such as prevention, mitigation and preparedness while the second includes the activities of response, recovery and rehabilitation.

The important characteristic of expand and contract model is that it can be analysed as a continuous process. The different disaster

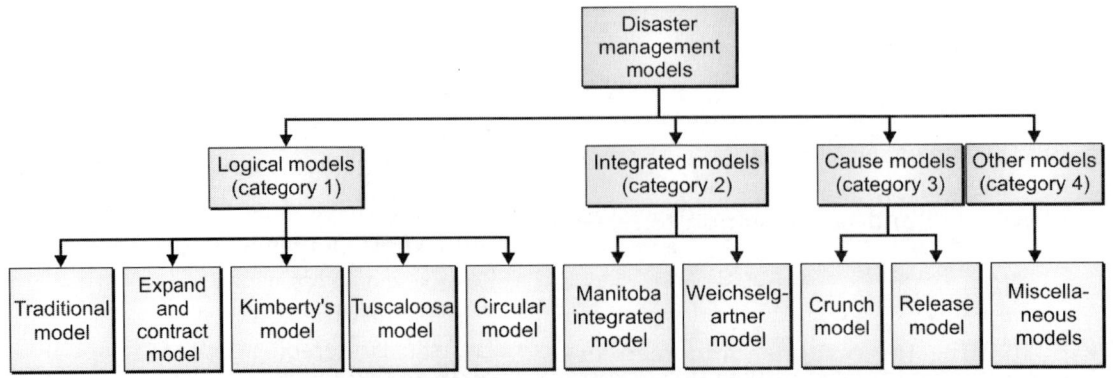

Fig. 3.5: Categorisation of disaster management models

management phases, rather than in a sequential manner, run parallel to each other, albeit with varying degrees of emphasis. These activities are expressed as the different strands (ADPC, 2000; Atmanand, 2003) and continue side by side, expanding or contracting as needed (DPLG-2, 1998). This model overcomes the limitations of the traditional model which is sequential in nature. This approach acknowledges that disaster management is a discipline which consists of various activities and actions that occur simultaneously.

Kimberly (2003) defines mitigation, preparation, response and recovery as four phases of disaster management. This model portrays response as the biggest and most visible phase of disaster management. It places mitigation and preparation at the base, suggesting that they are both driving forces behind a successful response. The recovery phase has been placed at the top because it is what remains after the response. Moreover, it takes the largest amount of time and is the most costly. The limitation of this model is that it is very much focused on emergency management in hospitals and cannot be significantly used in other applications. Since this model is restricted to hospital emergency management, its scope is limited.

Tuscaloosa emergency management model (2003), which is an open-ended process. The four phases in the cycle begin and end with mitigation that is, the on-going attempt to limit the effects of a disaster.

Kelly (1998) proposed the circular model for disaster management. It helps reduce the complexity of disasters and also handles the non-linear nature of disaster events. The model is more focused on practical disaster management needs than other disaster models. It lies in its ability to help in defining and elaborating the relationship between inputs and impacts rather than simply classifying disaster stages. The main characteristic of this model is its ability to learn from actual disasters.

An integrated disaster management model is a means of organising related activities to ensure their effective implementation. Four main components can be identified:

1. Hazard assessment
2. Risk management
3. Mitigation
4. Preparedness.

The first task in an integrated disaster management model is hazard assessment which provides the information necessary for the next phase, risk management. This result in decisions about the balance of mitigation and preparedness actions needed to address the risks (Manitoba Health Disaster Management, 2002). This model has altogether six independent elements such as a strategic plan, hazard assessment, risk management, mitigation, preparedness and monitoring and evaluation. Each element observes its own boundaries and involves its own set of activities and processes. These elements are dependent on each other in terms of providing support and can be further broken down into layers of sub-components. The advantage of this model is that it provides a balance between preparedness and flexibility in order to respond fluidly to the specific needs of disasters. Since this model provides the link between actions and events in disasters such links can be tight or loose. For example, it strongly links hazard and risk management activities but fails to provide a tight linkage between the four stages of disaster management which are important elements in a disaster management process.

The overall objectives of the Weichselgartner model (2001) are the assessment of possible damage and the planning of future actions to reduce this possible damage. It is argued that the assessment of vulnerability alone will not reduce natural hazards. Therefore, it is important that all measures taken are constantly reviewed and assessed. The model illustrates the process cycle and the integration of geographic place-based concepts in disaster management.

The crunch model provides the framework for understanding the causes of a disaster (ADPC, 2000; Bankoff, 2001; Heijmans, 2001;

Cannon, 2004; Marcus, 2005). The progression of vulnerability of a community is revealed and the underlying causes that fail to satisfy the demands of the people are identified. The model then goes onto estimate the dynamic pressure and unsafe conditions.

The pressure and release model (Blaikie et al, 1994; ADPC, 2000; Heijmans, 2001; Marcus, 2005) can be considered as the reverse of the crunch model. It indicates how the risk of disasters can be reduced by applying preventive and mitigation actions. It begins by addressing the underlying causes, and analysing the nature of hazards. This leads to safer conditions which help in order to prepare the community to deal disasters. The Indian Ocean tsunami and its impact on millions of people in the region demonstrate the high vulnerability of people in disaster situations when many existing predisposing factors are also in place (Blaikie et al, 2005).

Keller and Al-madhari (1996) proposed a model for the probabilistic prediction of disaster magnitude consequences and return period. As such it is particularly suitable for obtaining risk profiles.

Turner (1976) elaborated the sequence of events, which are the basis of development of a disaster. These stages are: (1) Notionally normal starting points, (2) incubation period, (3) precipitation event, (4) onset, (5) rescue and salvage, and (6) fully cultural readjustment.

Shrivastava (1992) proposed a model for industrial crisis through comparison of three crises: The Bhopal disaster, the Tylenol poisoning and the explosion of space shuttle challenger.

In summary, several models for disaster management have been proposed by researchers and agencies. The significance and usefulness of these different models have been discussed above, highlighting the instances and areas of applicability.

A Proposed Comprehensive Model for Disaster Management

Alexander (1997) argues that there is room for improvement in the approaches to disaster management based on the following three factors: (1) Death tolls have not fallen dramatically in response to improved mitigation, (2) large-scale transfer of technology has not occurred, and (3) disaster relief has not been adequately combined with mitigation and economic development. Therefore, this section proposes a comprehensive model for disaster management with improvements over existing models.

The models discussed in the previous section describe how the relationship between different phases of the disaster management process is mediated. It can be inferred from the study of these models that most revolve around the four major phases of disaster management. (i) Prevention, (ii) mitigation, (iii) response, and (iv) recovery. Such models are not planned to cover all the aspects of the disaster management domain and have some limitations; for example, logical models (category 1) do not go beyond describing disaster stages and only provide conceptual frameworks for the very basic activities of a disaster. The expand-contract model of category 1 does not encapsulate hazard assessment and risk management activities. Similarly, the crunch and release model only identifies the underlying causes of a disaster and do not highlight other major activities of disaster management.

The integrated model (category 2) covers most of the activities of the disaster management domain but does not encapsulate the activities of response and recovery. In addition, it only states the top level actions of disaster management rather than providing the detailed activities involved in each phase.

In category 3, the models focus on vulnerable conditions that might affect disaster management by identifying the underlying pressure and root causes of a disaster. The discussion about conditions affecting the disaster management cycle is limited to vulnerability.

The analysis of the abovementioned three categories reveals the following limitations:
• The design of most of the models revolves around the four main phases of disaster

management: (i) Prevention, (ii) mitigation, (iii) response, and (iv) recovery.

- There is no single model that encapsulates most of the major activities of disaster management within a single framework.
- The abovementioned models do not consider environmental conditions that might affect the severity of a disaster. They only think of environment as another disaster category.
- Some models fail to present a comprehensive description of disaster management activities within a single model. Furthermore, the arrangement of activities (if any) is not in a logical sequence.
- The evaluation and analysis of information and data related to a current disaster are highly important and essential ingredients in the mitigation of future disasters. The existing models do not give effective consideration to evaluation and analysis.

The current models, in terms of the three different categories, lack all of the required features and functionalities that would enable them to manage a disaster in a comprehensive manner. A comprehensive disaster management model, which supports different stages and phases of a disaster management cycle, can fill in the gap which occurs in the current models. In addition, such a model should have the ability to handle complex and difficult disaster scenarios which are not addressed by the current models.

Generally, in major disasters, various resources, conditions and activities are involved; identifying and utilising such resources, conditions and activities at a detailed level should be the goal of a disaster management model. Incorporating this level of activities and conditions affecting disasters, into existing models, would provide the basis for an effective, useful and practical disaster management model; one which would expand the attention to the full range of concerns about preparedness, mitigation, response and recovery.

Considering these limitations, and the insignificant highlighting of the conditions affecting disaster phases, the investigator present, a more comprehensive model (Fig. 3.6) which encapsulates all the required activities of disaster management.

The proposed comprehensive model is built upon linking the following: (1) Hazard assessment and risk management activities, (2) risk management activities and disaster management actions. The distinctive feature that it takes into account is the arrangement of activities in a logical sequence. It is applicable and based on a series of easy-to-determine factors which are combined in a simple way. The result of this combination and linkage of steps is a comprehensive disaster management model. The model is simple and intelligible; no expert knowledge is needed for its comprehension. Therefore, any technological based infrastructure (such as cyber infrastructure) can be linked to tackle disaster management problems. The model incorporates environmental conditions, which makes it possible to analyse and separate the environmental issues from a disaster.

Earlier in this section, the limitations of existing disaster management models were highlighted. Based on these, the possible improvements which have been incorporated in the proposed comprehensive model (conceptual):

1. The design of a comprehensive model does not revolve around four fundamental phases of disaster management. It has been segregated into six main components: Strategic planning, hazard assessment, risk management, disaster management actions (four fundamental phases of disaster management), monitoring and evaluation, and environmental effects.

2. Within the comprehensive model these six main components are further decomposed into various activities which are required in carrying out disaster management operations.

3. The disaster management actions are performed in a sequential manner in order to mitigate a disaster.

4. All disaster management measures, and actions taken, are constantly reviewed and

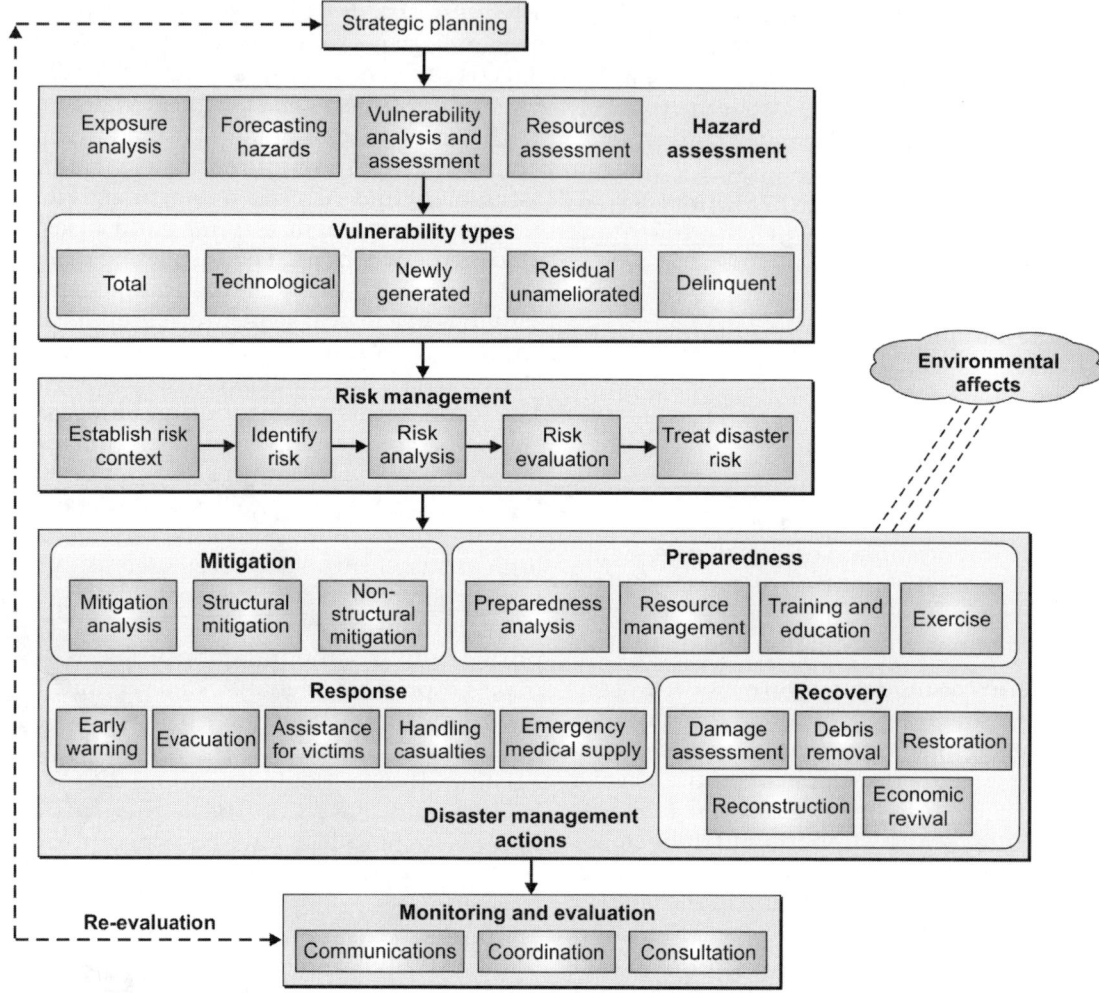

Fig. 3.6: Comprehensive model for disaster management by Alexander (1977)

assessed within the context of varying environmental conditions.

5. The results of, and assessments derived from the comprehensive model can be utilised as an input for a new evaluation which is obtained through the monitoring and evaluation module. Therefore, the evaluation of all measures, and feedback to the strategic planning module, is recommended.

6. The models discussed in the literature generally capture disaster management in a limited context, commonly revolving around mitigation, preparedness, response and recovery. But the proposed model extends this to include the changing effects of the environment in addition to other factors.

7. The assessment of possible disaster events is a very important issue when mitigating disasters. This important issue is addressed with the incorporation of hazard assessment and risk management modules in the comprehensive model. The risk management module in the comprehensive model is derived from the Australia, New Zealand Risk Management Standards (Salter, 1997; Standard-Australia, 1999).

8. Alexander (1991) proposed an approach to vulnerability assessment based on simple conceptual equations. Based on that approach, overall vulnerability can be broken down into a series of components based on different aspects of the problem. Therefore, Alexander (1997) suggested five heuristic classifications of types of vulnerability based on their societal context. The comprehensive model has adopted such classifications of vulnerability and incorporated them in the hazard assessment module.

The abovementioned improvements that have been incorporated in the comprehensive model suggest the following:

1. A large number of essential issues and activities are involved in the disaster management process.
2. This results in a highly complex system. The study of such activities and issues has raised inter-related problems associated with disaster management: The complexity and uncertain nature of the disaster management area is due to a large number of functions, features and activities.

Layered Relationship Derived from the Proposed Comprehensive Model

The analysis of the comprehensive model (Fig. 3.7) shows that it can be observed as a two-layered framework. The first layer shows the relationship between hazard assessment and risk management, the second highlights the relationship between the risk management and the disaster management actions which are mitigation, preparedness, response and recovery. The layered relationship is shown in Fig. 3.7.

In the literature, matrices have been used to represent different aspects of disaster management. For example, Kieft and Nur (2001) claimed that during disasters, the community's vulnerabilities are more pronounced than their capacities. To identify these, a capacity and vulnerability analysis matrix was drawn to examine various aspects.

Yasemin and Davis (1993) also produced a matrix which gives a rough indication of which actors might participate in the five stages of a disaster recovery phase. Salter (1997), suggested that within the risk management framework, the identification and analysis of risk concentrates on the interaction between "source of risk" and "element of risk". Therefore, Salter used a matrix approach to display such interactions within a disaster management context. Menoni (1996) also used a matrix-based approach to analyse the relationship between risk assessment and urban and regional planning. The present researchers have used a similar approach where present researchers perform an analysis of hazard assessment,

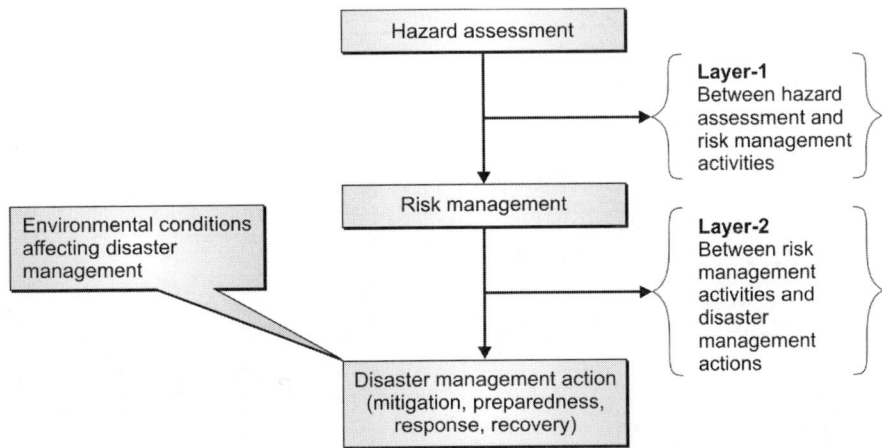

Fig. 3.7: Layered relationship derived from the proposed comprehensive model

risk management and disaster management actions. Researchers have incorporated the layered relationship drawn from disaster management activities in two matrices (Fig. 3.8):

1. *Activity matrix A:* Hazard assessment and risk analysis
2. *Activity matrix B:* Risk analysis and action (mitigation, preparedness, response and recovery)

According to McEntire (2002) and Auf der Heide (1989), social and behavioural research indicates that coordination is a major challenge among individuals, groups and agencies that respond to disasters. Therefore, the ability to communicate, coordinate and work effectively as a team can be a major factor in the success of any emergency plan. In response to these issues, researchers highlight the main problems associated with the development of disaster management systems. These are:

1. Establishing techniques for dynamic monitoring of disasters.
2. Failure in maintaining communication links.
3. The slow access to data which makes for poor updating of disaster-related information.
4. Difficulties in disaster-related data collection and integration.
5. Communication and collaboration among agencies.
6. Designing techniques for automated data processing from distributed sources.
7. Designing and developing decision support system to help emergency managers achieve effective decision-making for different disaster management activities such as mitigation, preparedness, response and relief.
8. Multiple models are required for decision-making.
9. Varying environmental affects which can significantly change the severity of a disaster.

Figure 3.9 shows problems associated with disaster management. As mentioned earlier, such problems arise due to the complexity involved in managing a large number of activities.

Disaster Knowledgeable Community Model

People's engagement in disaster management is widely believed to open new windows to public decision-making. Different countries

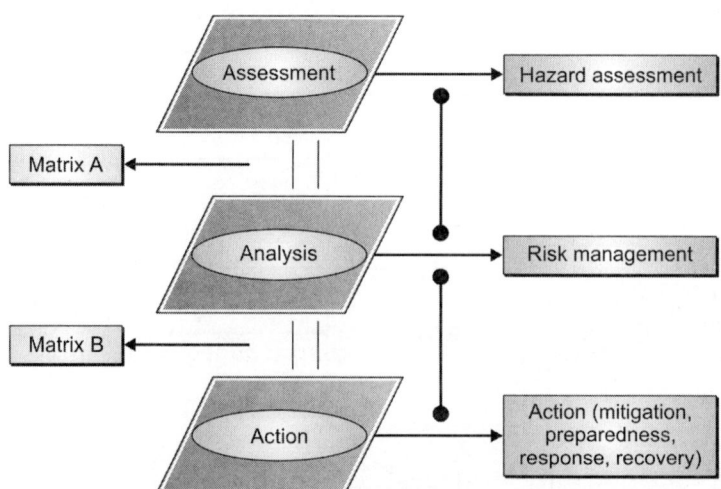

Fig. 3.8: Layered relationship and matrices

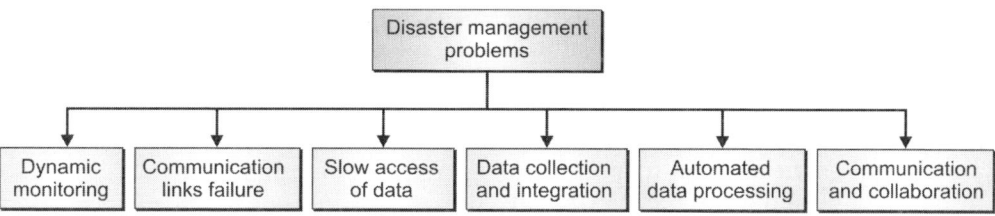

Fig. 3.9: The overall complexity of the disaster management domain

define people's engagement differently. For instance, people's engagement in Iran means involving the whole society in (responding to) disasters. In the United Kingdom and Australia, people's engagement may be defined as community engagement. The concept of community usually refers to people at the local level. For example, Buckle (2004) defines *community* as people at a local level who are not organised by emergency services but who have skills, resources, and capacities that allow them to provide services to people at risk in disasters. The notion of volunteerism is inherent in Buckle's definition of *community* engagement in disasters. Cultural differences among countries should, of course, be considered in understanding and learning from various earthquake experiences.

The Model of Disaster Knowledgeable Community in Practice

As Schneider (2002) notes, emergency management practitioners and community leaders may not always view the post-earthquake action in a wider framework. Apparently, the 'disaster knowledgeable community' model needs explanation to be translated into action. The metaphor of waves is used here to bridge the gap between the theory expounded in relevant literature and actual translation of the theory into practice. Observations of the early stage of the Bam earthquake constitute the practical experience leading to the metaphor of waves. There are several post-disaster waves: Physical, social, economic, political, and medical. 'Physical wave' is a force that not only destroys properties but also moves the other waves.

The physical wave is sometimes known as the disaster itself, particularly in natural disasters such as earthquakes and floods. Evidently, controlling the physical disaster wave in Iran could prevent the other disaster waves from being set in motion.

'Social wave' refers to the force that damages the structures and functions of a community's social order. For example, families as solid cells of community do not function normally. Specific social groups such as women, children, the elderly and the handicapped need special attention, and their numbers increase suddenly. Balali et al. (2004) note that the majority of disaster preparation plans are devised with the ordinary citizens in mind, while children, the disabled, pregnant women, and people with special needs are ignored. They suggest including these groups in mitigation programs. Maintaining communication is very important in disasters. The Bam experience proved the vulnerability of public communication systems, such as telephone lines, to severe disasters. Hesam and Mehrabi (2004) suggest that developing a limited personal communication system would be very helpful in practice. This would be a technology providing effective communication service for a small group of people such as members of a family, in crisis situations. The system should function even without electricity to connect members of a group or a family and also link any small group with the wider public systems.

'Economic wave' disturbs the normal and ordinary economic life. In this wave, distribution and supply of goods and services comes to a halt, losing its normality. Early response to this wave includes aid provided from new sources, usually outside the affected community. The main concern at this stage is equal distribution of goods and services in the short-term. In the long-term, funding becomes a serious problem for public officials. Damaged properties increase the level of poverty. Funding the reconstruction of properties becomes a long-term concern. 'Political wave' affects public officials and decision-makers at national, provincial, and local levels. This wave is sometimes very strong at the local level as it destroys local public entities and institutions.

Perhaps the real meaning of emergency is captured in the 'medical wave'. Rescuing is the vital function of disaster response at the impact stage. The medical wave affects medical systems in a vast area of the country. Coordinating medical teams within the affected area and the rest of the country was a complex task. This wave is less manageable from the top. Eshraghi (2004) states that the normal medical system is vulnerable to severe disasters. Eshraghi underlines the need for self-organising medical teams in the early days of severe disasters. Trained individuals with required equipment can render medical services to the victims early after the disasters. These individuals can also move and form medical teams in the affected areas.

There are two waves in severe disasters—one wave, demanding medical attention, moves from the centre of the disaster, while the other wave supplies and responds to medical needs of the affected area. These two waves move in opposite directions. The meeting of these waves may or may not be managed properly.

THEORIES OF DISASTER MANAGEMENT

Theory is a set of interrelated proposition that allow for the systematisation of knowledge, explanation and prediction of social life and generation of new research hypothesis.

A paradigm is a fundamental image of the subject matter within a science. There will not be an over-arching or meta theory of Disaster Management. While it would no doubt be difficult for any single concept to capture every conceivable variable and issue pertinent to disasters, it is clear that some theories are more inclusive than others. Some important theories (known as paradigm) from Fig. 3.10 helps us to know why some problem is occurring and, more importantly, how it can be resolved or mitigated.

The radical and cultural/institutional theories presented by Marx and Weber have had a profound impact on disaster. On the one hand, poverty is a major causal explanation of disaster. It advocates a restructuring of social, political and economic relations so that calamities can be reduced which is reflected in Fig. 3.11).

It has been well documented in the Table 3.2 that economic conditions and political powerless are related to disaster vulnerability. Poor people, minorities and other marginalised people are most likely to live in dangerous areas and substandard housing, and are least able to deal with the adverse effects of droughts, tornodoes and other hazards.

When a disaster occurs, people and resources will flow to the scene and new organisations will appear almost instantaneously.

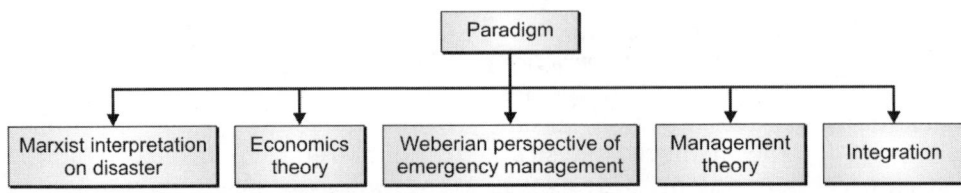

Fig. 3.10: Theories of disaster management

Marxist Interpretation on Disaster

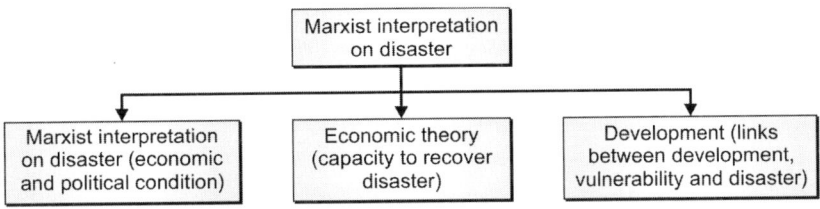

Fig. 3.11: Marxist interpretation on disaster

Table 3.2: Marxist interpretation on disaster (integration of theories)

Theories	Cause (in disaster vulnerability) of the approach	Effect of the approach
1. Marxist interpretation on disaster (economic and political condition)	Economic conditions and political powerlessness are related to DV	Poor, minorities and other marginalised people are most likely to live in dangerous areas and affected by disasters
2. Economic theory (capacity to recover disaster)	Economic prosperity and ability to economically recovery	Quickly recover, capacity to manage disaster, assistance to face disaster that hits in society, nations and country
3. Development (links between development, vulnerability and disaster)	Well-planned or haphazardly development cause less or more vulnerability	Disaster result may be increased or reduced

From Fig. 3.12, it is seen that first task to respond any kinds of disaster is to build up capacity and trained the villagers. If development is occurs haphazardly, vulnerability will increased and additional disasters will result. It is also seen from the Fig. 3.12 that environmental degradation may create additional vulnerability in future. Modern communications equipment as well as advanced hardware and software applications may build our capabilities to prevent, prepare for and respond to disasters.

Economic Theory (Capacity to Discover Disaster)

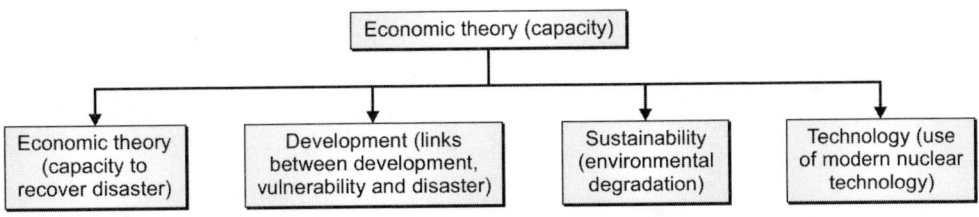

Fig. 3.12: Coping mechanism of disaster

Table 3.3: Management theories

Theories	Cause (in disaster vulnerability) of the approach	Effect of the approach
1. Management theory (political and organisational actions)	Effective or ineffective leadership and strategic planning may be reducing or increasing vulnerability	The ability of emergency managers to sway public opinion and actively pursue objectives will likely increase steps taken for mitigation and enhance the preparedness level of the jurisdiction
2. Decision theory (availability of information)	Lack of information may make responders and citizens vulnerable to injury, death, disruption and other adverse effects of disasters	Incorrect perceptions, bureaucratic politics and factors consequently have a bearing on the creation of risk, susceptibility, resistance and resilience
3. Systems theory (socioeconomic and organisational system)	Natural, built, technological, social, political, economic, cultural, organisational and psychological systems have direct impact on level of vulnerability. Diverse systems interact in complicated ways	The location and construction of our buildings may be due to policy enforcement, cultural preferences, income levels, risk perception, urbanization, education of the population, etc.
4. Chaos theory (systems theory's many variables)	Many variables that interact to produce vulnerability	Vulnerability may be reduced by addressing multiple variables simultaneously
5. Paper plan syndrome (emergency operations plan)	Paper plan syndrome without developing capabilities to implement the plan does nothing to ensure adequate response and recovery operations	Vulnerability can only be addressed through risk assessments, land-use planning, improved engineering
6. Compliance model of evacuation (a number of variables)	These variables would help responders predict who is not likely to leave and determine what can be done to encourage them to evacuate	Compliance model influences whether or not a person evacuates during an impending disaster
7. Policy making (political processes)	Political processes allocate both knowing and unknowing values in society that distribute vulnerability among the population	Through policy, citizens become more concerned about common forms of vulnerability; if our policy only takes into account expert advice, our society may become more vulnerable to civil hazards

Management Theory (Political and Organisational Activities)

Management theory is a combination of seven theories which is seen in Fig. 3.13.

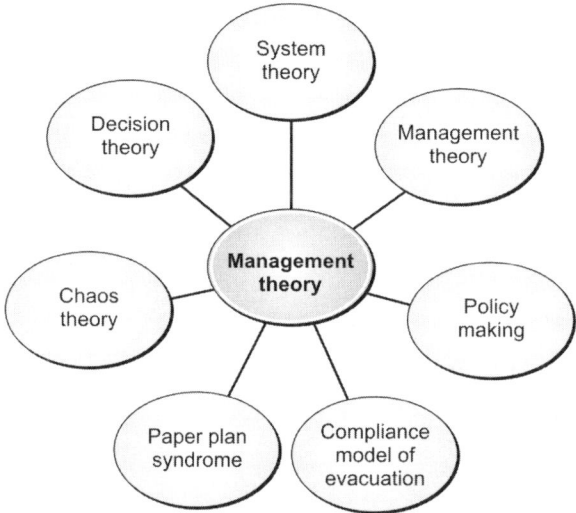

Fig. 3.13: Graphical presentation of management theory

Management theory: Disasters are political and organisational problems. Some of the vulnerability in our communities may be corrected through effective leadership and strategic planning. The ability of emergency managers to sway public opinion and actively pursue objectives will likely increase steps taken for mitigation and enhance the preparedness level of the jurisdiction (thereby reducing vulnerability).

Decision theory: From Table 3.3, disasters are almost always characterised by a lack of information. It is this uncertainty that makes responders and citizens vulnerable to injury, death, disruption and other adverse effects of disasters. Incorrect perceptions, bureaucratic polities and factors consequently have a bearing on the creation or reduction of risk, susceptibility, resistance and resilience.

Systems theory: Table 3.3 illustrates that the national, built technological, social, political, economic, cultural, organisational and psychological environments have direct impact on our level of vulnerability. But these diverse systems interact in complicated ways. For instance, the location and construction of our buildings may be due to policy enforcement, cultural preferences, income levels, risk perception, urbanisation education of the population, etc. systems theory is thus applicable to the model of vulnerability presented in this paper.

Chaos theory: As can be seen in Table 3.3 that there are many variables that interact to produce vulnerability. Chaos theory suggests that it is impossible to detect simple linear cause and effect relationship. Instead, there are many variables that interact in convoluted ways to produce disaster. Chaos theory would thus recommend that vulnerability be reduced by addressing multiple variables simultaneously (i.e. there may be order in policies that appear to be chaotic).

Paper plan syndrome: Some communities assume that the presence of an emergency operations plan is all that is needed to deal with disasters. Having a written document without developing capabilities to implement the plan does nothing to ensure adequate response and recovery operations. In addition, other types of vulnerabilities can only be addressed through risk assessments, land-use planning, improved engineering: they planning, improved engineering; they are not amenable to emergency planning alone.

Compliance model of evacuation: Table 3.3 indicated that there are a number of variables that influence whether or not a person evacuates during an impending disaster (e.g. age, gender, race, education, activities of neighbours, etc). An understanding of these variables would help responders predict who is not likely to leave (thus increasing their vulnerability) and determine what can be done to encourage them to evacuate (thus reducing their vulnerability.

Policy making: Political processes not only allocate values in society, they also knowingly or unknowingly distribute vulnerability among the population. If we institute law based on public opinion, we may neglect less visible but more common forms of vulnerability (e.g. citizens are more concerned today about our vulnerability to terrorism than our vulnerability to flooding). On the other hand, if our policy only takes into account expert advice, our society may become more vulnerable to civil hazards such as riots and terrorism (e.g. the public may become hostile if its desires are not considered in the policy process).

Integration Theory

Because there are so many participants involved in emergency management, it is crucial that these organisations espouse and implement consistent and achievable policies. Figure 3.14 depicts that the integration theory has three segments. These are integration, networking and collaboration and preparedness and improvisation. The current national disaster assistance program discourages local communities from reducing vulnerability because federal resources can be acquired for relief and reconstruction purposes (Berke, 1995). The delegation of additional homeland security tasks to the local level without the provision of sufficient human resources may also increase the vulnerability of jurisdictions to disasters; emergency managers may not be able to meet all of the assignments and responsibilities that need to be performed before disaster strikes. Vulnerability is accordingly related to the concept of integration also.

Networking and collaboration: Disasters require multi-organisational responses. Getting public, private and non-profit agencies to work together before, during and after disasters are likely to reduce liabilities and raise capabilities. For instance, from Table 3.4, it encouraging developers and landowners to support safe development will reduce liabilities while increased contact and cooperation among disaster response organisations will build capabilities.

Preparedness and improvisation: It is possible to see ties between vulnerability and the twin foundations of emergency management also. What emergency managers and first responders do before and after a disaster

Integration (national, international and local)

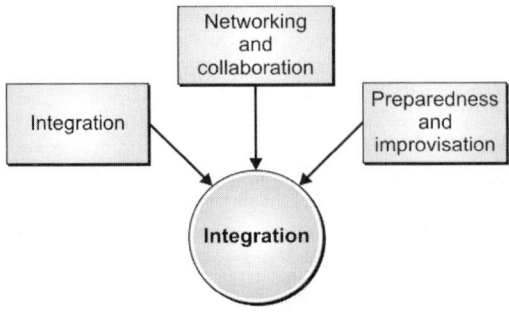

Fig. 3.14: Graphical presentation of integration theory

Table 3.4: Integration (national, international and local)

Theories	Cause (in disaster vulnerability) of the approach	Effect of the approach
1. Integration (national, international and local integration)	Emergency managers may not be able to meet all the assignments and responsibilities	The delegation of additional homeland security tasks to the local level without the provision of sufficient human resources may not be able to meet all the jurisdiction to disasters
2. Networking and collaboration (multi-organisational response)	Getting public, private and non-profit agencies to work together before, during and after disasters are likely to reduce liabilities and raise capabilities	Networking and collaboration increases contact and cooperation among disaster response organizations that builds capabilities
3. Preparedness and improvisation (ties between vulnerability and emergency management)	Planning, training and exercising are developing capabilities of the community for managing disaster	As part of preparedness, improvisation may either make people susceptible or may increase opportunities for resilience

has a great bearing on vulnerability. If planning, training and exercises are taken seriously, the capability of the community is enhanced (Table 3.4). Improvisation, for its part, may either make people susceptible or may increase opportunities for resilience.

Questions

1. Define disaster management.
2. Define post-disaster stages.
3. Briefly discuss the Disaster Management Cycle with suitable examples.
4. What is the disaster management training strategy and/or plan of the country with regards to disaster management?
5. What is the structure/organogram for disaster management in the country? What is the number of personnel in each organization/agency?
6. Discuss the paradigm shift in disaster management in India.
7. What are the functions of the provincial and local authorities in disaster management?
8. How are various sectors such as (agriculture, health, infrastructure, education, water resources, interior) engaged in risk management issues?
9. What are the functions of the ISDR?
10. What is NDMA? Explain its role in disaster management.
11. What organization is responsible for disaster warning?
12. How are the warnings transmitted to officials and to the population? Channels of dissemination communication.
13. Write about state and international level disaster management programmers.
14. Write the role played by Armed forces, NDRF police and civel defence to prevent disasters.
15. How the man-made hazards controlled and regulated? Discuss.
16. Discuss the different aspects of disaster mitigation through advanced technology.
17. Write short notes on emergency stage in disaster management.
18. Define the term rehabilitation.
19. List out the role of media in disaster
20. Explain the various strategies to be adopted for the disaster preparedness.
21. What is "financial arrangement"?
22. What do you mean by Structural and Non-Structural Mitigation for Disaster Management?

23. What is the role of Non-government and Inter-governmental Agencies in Disaster Management?

24. List some methods of effective dissemination of information required for disaster management.

25. What do you understand by rehabilitation?

26. Explain International Decade for National Disaster Reduction (IDNDR).

27. Discuss in brief the Disaster Management Act, 2005.

28. Discuss the concept and significance of coordination in disaster management.

29. Discuss the role of SDMA and DDMA.

30. What are the major phases of electronic warning system?

31. Explain various community-based disaster management.

32. What are the theories of disaster management?

Bibliography

ADPC (2000), "Community Based Disaster Management (CBDM): Trainer's Guide, Module 4: Disaster Management". Asian Disaster Preparedness Center (ADPC). Bangkok, Thailand.

Alexander D. "Applied Geomorphology and Impacts of Natural Hazards on the Built Environment", *Natural Hazards* 4(1):1991; 57-80.

Alexander D. "The Study of Natural Disasters, 1977–1997: Some Reflections on a Changing Field of Knowledge". *Disasters* 1997; 21(4):284–304.

Md. Abu Bakar Shamim (2014). Theoretical approaches of disaster management.

Alexander D. Natural Disasters, London: UCL Press. 1993; p. 631.

Allen K, Green S, Zubrow E. (eds) (1990). Interpreting Space. GIS and Archaeology. London: Taylor & Francis.

Anders Levermann, Peter U. Clark, Ben Marzeion, Glenn A. Milne, David Pollard, Valentina Radic, and Alexander Robinson. "The multimillennial sea-level commitment of global warming". PNAS. 13 June 2013; 110:13745–13750.

Andino, Jean M. (October 21, 1999). "Chlorofluorocarbons (CFCs) are heavier than air, so how do scientists suppose that these chemicals reach the altitude of the ozone layer to adversely affect it?". Scientific American.

Ashmore, Wendy and Knapp B. A. (eds) (1999). Archaeologies of Landscape. Contemporary Perspectives. Massachussets, U.S.A. & Oxford, London: Blackwell Publishers.

Atmanand. "Insurance and Disaster Management: The Indian Context" *Disaster Prevention and Management* 2003, 12(4):286–304.

Auf der Heide E. (1989), *Disaster Response: Principles of Preparation and Coordination*. C. V. Mosby Company, Toronto.

Balali MR, Davoodi M, Rasekhjam A, Navidi A. 2004, 'Plans and Requirements for Disasters', *The Second International Congress of Disaster Management and Health Proceedings*, Shokravi Publication, Tehran.

Bankoff G. (2001), "Rendering the world unsafe: 'Vulnerability' as Western Discourse". *Proceedings of International Work-Conference on Vulnerability in Disaster Theory and Practice*, Wageningen, 29/30 June, 2001.

Berke PR (1995). "Natural Hazard Reduction and Sustainable Development: A Global Reassessment". Working paper no. 595-02 Center for Urban and Regional Studies.

Barrows HH. 'Geography as human ecology' in Annals of the association of the American Geographers 1923; 13:1–14.

Bhargava, Renu. "Earth, Nature, Environment, Ecosystem and the Human Society" Journal of Multidisciplinary Studies, 2011; 1(1).

Blaikie P., Cannon T., Davis I. and Wisner B. (1994). *At Risk: Natural Hazards, People's Vulnerability and Disasters*. Routledge, London.

Blaikie P., Mainka S. and McNeely J. (2005). "The Indian Ocean Tsunami Reducing Risk and Vulnerability of Future Natural Disasters and Loss of Ecosystem Services". *An Information Paper: The World Conservation Union (IUCN)*, Switzerland, Feb. 2005.

Broad, William J. (2010-06-02). "Nuclear Option on Gulf Oil Spill? No Way, U.S. Says". New York Times. Archived from the original on 2010-06-17. Retrieved 2010-06-18.

Buckle, P 2004, 'A Comparative Assessment of Community based Recovery Management in England and Australia', Coventry Center for Disaster Management, Coventry University.

Business Continuity Institute (2013). Good Practice Guidelines: Global Edition. BCI.

Cannon T. (2004), "At Risk: Natural Hazards, People's Vulnerability and Disasters".

Proceedings of the CENAT Conference, Switzerland 28 November – 3 December, 2004.

Climate Change 2014 Synthesis Report Fifth Assessment Report, AR5 (Report). Intergovernmental Panel on Climate Change. 2014.

Cuny, Fred. Disasters and Development. New York and Oxford: Oxford University Press, 1983.

Czekaj, Laura. "Ready for the World: Paramedics Train for International Disasters", Ottawa Sun, 5 November 2006.

Daly, Herman E., 1980, Economics, Ecology, Ethics: Essays toward a steady-state economy NY: W.H.Freeman & Co.

Dave, R.K. (2009). Role of Media in Disaster Management. National Disaster Management Authority, Ploicy and Plan.

Disaster Definitions, The Johns Hopkins and the International Federation of Red Cross and Red Crescent Societies.

DPLG-2 (1998), "Green Paper on Disaster Management: Chapters 2 and 3". Available: http://www.local.gov.za/DCD/policydocs/gpdm/gpdm2–3.html. Accessed: December 2004.

Dr Renu Bhargava, Bhatter College, Journal of Multidisciplinary Studies (ISSN 2249–3301), "Volume 1(1). 2011".

Drabek, Thomas E. (1986). Human System Response to Disaster: An Inventory of Sociological Findings. London: Springer-Veriag.

Emergency Response and Recovery, Cabinet Office, 29 October 2013.

EPA,OA, US. "Myths vs. Facts: Denial of Petitions for Reconsideration of the Endangerment and Cause or Contribute Findings for Greenhouse Gases under Section 202(a) of the Clean Air Act | US EPA". US EPA. Retrieved 2017-08-07. The U.S. Global Change Research Program, the National Academy of Sciences, and the Intergovernmental Panel on Climate Change (IPCC) have each independently concluded that warming of the climate system in recent decades is "unequivocal." This conclusion is not drawn from any one source of data but is based on multiple lines of evidence, including three worldwide temperature datasets showing nearly identical warming trends as well as numerous other independent indicators of global warming (e.g., rising sea levels, shrinking Arctic sea ice).

Eshraghi H R (2004). 'Medical Group Organizing and Management in Disasters'. *The Second International Congress of Disaster Management and Health Proceedings*, Shokravi Publication, Tehran.

Federal and Military (2012). Fire & EMS, Law Enforcement/What is the Role of Police, Fire and EMS After a Natural Disaster Strikes?

Fischlin; et al., "Section 4.4.9: Oceans and shallow seas—Impacts", *in* IPCC AR4 WG2 2007, Chapter 4: Ecosystems, their Properties, Goods and Services, p. 234.

Fowler, John, 1986, Energy and the Environment, NY: McGraw Hill.

Frederick Krimgold. The Role of International Aid for Pre-disaster Planning in Developing Countries, Audelringen for Arkitektur, KTH Stockholm, 1974, p. 65.

Frich A, Alexander LV, Della-Marta P, Gleason B, Haylock M, Klein Tank AMG, Peterson T. "Observed coherent changes in climatic extremes during the second half of the twentieth century" (PDF). Climate Research. January, 2002, 19:193–212.

Bankoff G, Frerks G, Hilhorst D (eds) (2003). Mapping Vulnerability: Disasters, Development and People. ISBN 1-85383-964-7.

Gillis, Justin (2015-11-28). "Short Answers to Hard Questions About Climate Change". The New York Times. ISSN 0362-4331. Retrieved 2017-08-07.

Global and regional sea level rise scenarios for the United States (PDF) (Report), (NOAA Technical Report NOS CO-OPS 083 ed.). National Oceanic and Atmospheric Administration. January 2017. Retrieved 25 January 2017.

Goel, S. L. (2006). Policy and Administration. Deep and Deep Publications, New Delhi, p. 404.

GOI-UNDP (2002–2009). Disaster Risk Management Programme, New Delhi, p. 21.

Guha, Ramachandra, 2000 Environmentalism, A Global History, New Delhi, Oxford University Press, Lomborg, The Skeptical Environmentalist.

Hartmann DL, Klein Tank AMG. Rusticucci M (2013). "Two Observations: Atmosphere and Surface" (PDF). IPCC WGI AR5 (Report). p. 198. Evidence for a warming world comes from multiple independent climate indicators, from high up in the atmosphere to the depths of the oceans. They include changes in surface, atmospheric and oceanic temperatures; glaciers; snow cover; sea ice; sea level and atmospheric water vapour. Scientists from all over the world have independently verified this evidence many times.

Heijmans A. (2001), "Vulnerability: A Matter of Perception", Disaster Management Working Paper 4/2001, Benfield Greig Hazard Research Centre University College of London.

Hesam S and Mehrabi F (2004), 'Personal Limited Communication System'. *The Second International Congress of Disaster Management and Health Proceedings*, Shokravi Publication, Tehran.

High Power Committee on Disaster Management (2001). Report of the High Power Committee on Disaster Management, National Center of Disaster Management, New Delhi.

Hobbs, Peter V, Radke, Lawrence, F. "Airborne studies of the smoke from the Kuwait Oil Fires". Science. 1992; 256(5059):987–991.

http://environment.nationalgeographic.com/ environment/natural-disasters/avalanche-profile/

http://science.howstuffworks.com/nature/ natural-disasters/avalanche.htm Ibid pp. 44–56.

http://www.unisdr.org/files/7817_ UNISDRTerminologyEnglish.pdf

IFRC (2011). Guidelines for Emergency Assessment. Geneva: International Federation of the Red Cross and Red Crescent Societies. http:// www.proventionconsortium.org/files/ tools_CRA/IFRC-guidelinesassessments-

IMD (2008). Management of Cyclone: National Disaster Management Guidelines.

IMD (2009). Site Report – Cyclone Aila.

ISDR (2000). Disaster Prevention, Education and Youth. Terminology on Disaster Risk Reduction. Geneva: United Nations.

ISDR (2001). Natural disasters and sustainable development: Understanding the links between development, environment, and natural disasters. Background Paper No. 5, Commission on Sustainable Development, Second Preparatory Session, 28 January–8 February 2002 (DESA/DSD/PC2/BP5), 10 pp.

J. Eugene Haas, Robert W. Kates, Martyn J. Bowden, Reconstruction Following Disasters, MIT Press, Cambridge, Massachusetts, 1977.

Keller AZ, Al-madhari AF. "Risk Management and Disasters". *Disaster Prevention and Management* 1996; 5(5):19–22.

Kelly C. "Simplifying Disasters: Developing a model for Complex Non-linear Events".

Proceedings of International Conference on Disaster Management: Crisis and Opportunity: Hazard Management and Disaster Preparedness in Australasia and the Pacific Region, Cairns, Queensland, Australia, 1–4 November, 1998; 25–28.

Kieft J. and Nur A. (2001), "Community-Based Disaster Management: a response to increased Risks to Disaster with Emphasis on Forest Fires". Available: www.fao.org/docrep/005/ AC798E/ac798e0e.htm. Accessed: April 2005.

Kimberly A. "Disaster Preparedness in Virginia Hospital Center, Arlington, after Sept 11, 2001" *Disaster Management and Response* 2003; 1(3):80–86.

Krajick, Kevin (2005). "Fire in the Hole". Smithsonian Magazine, 5(1). p. 54ff. Retrieved 2007-01-16.

Manitoba Health Disaster Management (2002), "Disaster Management Model for the Health Sector: Guideline for Program Development". Version 1, November 2002.

Marx K. (1967) Capital: Critique of political Economy, International Publishers, New York, NY.

Marcus O. (2005), "A Conceptual Framework for Risk Reduction". *World Conference of Disaster Reduction*, Kobe, Japan, 18–22 January 2005.

McEntire D. "Coordinating multi-organizational responses to disaster: lessons from the March 28, 2000, Fort Worth tornado". *Journal of Disaster Prevention and Management* 2002; 11(5):369–379.

McEntire AD (2004). The status of Emergency Management Theory: Issues Barriers and Recommendations for Improved Scholarship, Paper Presented at the FEMA Higher Education Conference, June 8, 2004, Emmitsburg, MD.

Meehl, George A.; Tebaldi, Claudia. "More Intense, More Frequent, and Longer Lasting Heat Waves in the 21st Century". Science. 13 August 2004; 305(5686):994–997.

Menoni S. "An Attempt to Link Risk Assessment with Land Use Planning: A Recent Experience in Italy" *Disaster Prevention and Management* 1996; 5(1):6–21.

Mondal, Debabrata, Sarthak Chowdhury, and Debabrata Basu. "Role of panchayat (local self-government) in managing disaster in terms of reconstruction, crop protection, livestock management and health and sanitation measures." Natural Hazards 94.1 (2018):371–383.

Mondal, Debabrata, Sarthak Chowdhury, and Debabrata Basu. "The role of gram panchayats in disaster management: A study in Aila affected areas in West Bengal." Indian Research Journal of Extension Education 14.3 (2016):51–54.

National Academy of Agricultural Science (2004). Disaster Management in Agriculture, Policy paper, 27.

National disaster management guideline (2015). Role of NGOs in disaster management.

National Disaster Response Force. Ministry of Home Affairs. 10 July 2013.

National Research Council (2006). Facing hazards and disasters: Understanding human dimensions. Washington, D.C.: National Academies Press.

NDMA: National Disaster Management Guidelines: Management of Floods. New Delhi: National Disaster Management Authority, Government of India, 2008; pp. 89–90.

NDMA (2009). Policy and Guidelines. New Delhi: National Disaster Management Authority, Government of India. Accessed at http://ndma.gov.in on April 27, 2005.

NFPA (2004). Standard on Disaster/Emergency Management and Business Continuity Programs. (http:www.nfpa.org/PDF/nfpa1600.pdf?sic=nfpa).

NFPA (2004). Standard on Disaster/Emergency Management and Business Continuity Programs. (http:www.nfpa.org/PDF/nfpa1600.pdf?sic=nfpa).

Noerdlinger PD, Brower KR. The melting of floating ice raises the ocean level. Geophysical Journal International, 2007; 170(1) 145–150.

Nordyke MD. (2000-09-01). "Extinguishing Runaway Gas Well Fires". The Soviet Program for Peaceful Uses of Nuclear Explosions (PDF). Lawrence Livermore National Laboratory. pp. 34–35. doi:10.2172/793554. Report no.: UCRL-ID-124410 Rev 2. Archived (PDF) from the original on 2016-12-23. U. S. Department of Energy contract no.: W-7405-Eng48.

Pan American Health Organization, A Guide to Emergency Health Management after Natural Disaster, Pan American Health Organization, Washington, D.C., 1981; pp. 3–4.

Public Health Guide for Emergencies, Johns Hopkins Bloomberg School of Public Health, 2013.

Read more on Brainly.in - https://brainly.in/question/131413#readmore

Rein G. Smouldering Fires and Natural Fuels. In CM Belcher et al (Eds). Fire Phenomena and the Earth System: An Interdisciplinary Guide to Fire Science. Wiley and Sons. 2013; pp. 15–34.

"Functions and Responsibilities". National Disaster Management Authority. Retrieved 28 October 2014.

"Global warming—definition of global warming in English | Oxford Dictionaries". Oxford Dictionaries | English. Retrieved 2017-08-07.

"IPCC, Climate Change 2013: The Physical Science Basis—Summary for Policymakers (AR5 WG1)" (PDF). p. 4. Warming of the climate system is unequivocal, and since the 1950s, many of the observed changes are unprecedented over decades to millennia.

"The Antarctic Ozone Hole Will Recover – June 4, 2015". NASA. Retrieved 2017-08-05.

"Twenty Questions and Answers About the Ozone Layer". Scientific Assessment of Ozone Depletion: 2010 (PDF). World Meteorological Organization. 2011. Retrieved March 13, 2015.

Robinson, Peter J. "On the Definition of a Heat Wave". Journal of Applied Meteorology. American Meteorological Society. April 2001, 40(4):762–775.

Safari M, Ghazanfari Z & Eskandari F 2004, 'Coordinating Manpower with Medical Personnel in Disaster', The Second International Congress of Disaster Management and Health Proceedings, Shokravi Publication, Tehran.

Salter J. "Risk Management in a Disaster Management Context" Journal of Contingencies and Crisis Management 1997; 5(1):60–65.

Schneid D, Thomas, Collins Larry (2001). Disaster Management and Preparedness. D. S. Thomas. New York, Lewis Publisher, NY.

Shennan I. (2013). Sea Level Studies: Overview. In: Elias SA, Mock J (eds). Encyclopedia of Quaternary Science (Second Edition). Elsevier, Amsterdam, Netherlands, pp. 369–376. ISBN 978-0-444-53643-3

Shi, Hanchang. "Industrial Wastewater: Types, Amounts and Effects." Point Sources of Pollution: Local Effects and its Control. I. UNESCO, Encyclopedia of Life Support Systems (EOLSS), 2002. 1–6. Web.

Son, Jeongwook; Aziz, Zeeshan; Feniosky Peña-Mora. Structural Survey, 2007; 26(5):411–425.

Song Z, Kuenzer C. "Coal fires in China over the last decade: A comprehensive review". International Journal of Coal Geology. 133:72–99.

Standard-Australia (1999), "AS/NZA 4360:1999 Risk Management, Standards Australia". Australia New Zealand standard, Homebush, Standards New Zealand, Wellington.

Turner BA. "The organizational and Interorganizational Development of Disasters" *Administrative Science Quarterly* 1976; 21:379–397.

Tuscaloosa EMA (2003). "Tuscaloosa County Emergency Management Cycle". Available: www.tuscoema.org/cycle.html. Accessed: April 2005.

U.S. Billion-Dollar Weather and Climate Disasters.

UN/ISDR (2004). Living with Risk: A Global Review of Disaster Reduction Initiatives. Geneva: United Nations Inter-Agency Secretariat of the International Strategy for Disaster Reduction (UN/ISDR). Available at http://www.unisdr.org/eng/about_isdr/basic_docs/LwR2004.

Warfield, C. (2005). The Disaster Management Cycle. Accessed on 23/01/08 at: http://www.gdrc.org/uem/disasters/1-dm_cycle.html.

Weber M (1965). Politics as vocation, Fortrees Press, Philadelphia, PA.

Weichselgartner J. "Disaster Mitigation: The Concept of Vulnerability Revisited" *Disaster Prevention and Management* 2001; 10(2):85–94.

White, G.F. (1945). Human adjustment to floods: A geographical approach to the flood problem in the United States, Chicago: Department of Geography, University of Chicago.

WMO (1989). Natural Disaster Reduction: How Meteorological and Hydrological Services Can Help by D.K. Smith, Switzerland.

www.fao.org/ranger service/

www.library.thinkquest.org

Yasemin A. and Davis I. (1993), "Rehabilitation and Reconstruction". Disaster Management Training Programme (UNDP).

Index

93

Reader's Notes

Reader's Notes

Reader's Notes

Reader's Notes

Reader's Notes